H. H. Prell · P. R. Day   *Plant-Fungal Pathogen Interaction*

Springer

Berlin
Heidelberg
New York
Barcelona
Hong Kong
London
Milan
Paris
Singapore
Tokyo

Hermann H. Prell · Peter R. Day

Plant-Fungal Pathogen Interaction

A Classical and Molecular View

With 15 Figures and 12 Tables

 Springer

Professor em. Dr. HERMANN H. PRELL
Institut für Pflanzenpathologie
und Pflanzenschutz
Universität Göttingen
Grisebachstraße 6
37077 Göttingen, Germany

Professor Dr. PETER DAY
Rutgers, The State University of New Jersey
Biotechnology Center for Agriculture
and the Environment
Foran Hall, Cook College
59 Dudley Road
New Brunswick, NJ 08901-8520, USA

Title of the original German edition. Hermann H. Prell: Interaktionen von Pflanzen
und phytopathogenen Pilzen. ISBN 3-437-35028-5; © Gustav Fischer Verlag, Jena 1996

Cover illustration: Telium with teleutospores of *Puccinia graminis* on a leaf. – Institute for
Plant Pathology and Plant Protection, University of Göttingen, Dr. Helga Nirenberg.

ISBN 3-540-66727-X Springer-Verlag Berlin Heidelberg New York

Die Deutsche Bibliothek – CIP-Einheitsaufnahme
Prell, Hermann H.:
Plant fungal pathogen interaction : a classical and molecular view /
Hermann H. Prell; Peter R. Day. – Heidelberg; New York;
Barcelona; Hong Kong; London; Milan; Paris; Singapore; Tokyo:
Springer, 2000
   Einheitssacht.: Interaktionen von Pflanzen und phytopathogenen Pilzen <engl.>
   ISBN 3-540-66727-X

Springer-Verlag Berlin Heidelberg New York
ein Unternehmen der BertelsmannSpringer Science + Business Media GmbH

© Springer-Verlag Berlin Heidelberg 2001
Printed in Germany

Cover design: Design & Production, Heidelberg
Typesetting: Mitterweger & Partner GmbH, Plankstadt
SPIN 10569959 31/3130 – 5 4 3 2 1 0 – Printed on acid-free paper

Preface

This book has been written for scientists and graduate students in plant pathology, biology, biochemistry, genetics and molecular genetics, for practitioners in plant protection and plant breeding, and for the interested layman. We have tried to produce a concise survey of the diversity of interactions between plants and phytopathogenic fungi. Our main objective has been to help the reader to comprehend the complexity of these interactions and the many different approaches used to analyse them. Some recent developments are also presented to demonstrate the close connections between plant pathology and basic research in biology.

Within the last 20 years many exhaustive reviews have been published, that deal with particular aspects of plant-pathogen interactions. However, it is difficult to obtain from them a coherent picture comprising all or most of their important features. Also the technical terms employed in the literature may be unfamiliar to many readers and definitions are often difficult to find. This book provides definitions, attempts to bring together the recent literature on host-parasite interactions, and discusses the range of experimental approaches that have been used to explore them.

We have relied fairly heavily on the review papers listed at the end of each chapter. More recent and particularly interesting results from original papers are also cited in the text and the references given at the ends of chapters under "relevant papers". Since we wished to keep relevant references together at the end of each chapter some references are repeated in other chapters.

The first edition of this book appeared in German in the Fischer-Verlag. (Prell, Hermann, H.: Interaktionen von Pflanzen und phytopathogenen Pilzen: Parasitierung, und Resistenz, Genetik und molekulare Phytopathologie. – Jena; Stuttgart: G. Fischer, 1996). This English translation of the German book is a revised and updated version of the German publication. Translation and improvements of this second edition were prepared by both authors, H.P. and P.D.. Both appreciate the revision of the first English draft of H.P. carried out by the Springer-Verlag and the publisher's patience in waiting for the complete manuscript. – H.P. appreciates very much the support given by Dr. B. Koopmann, Inst. for Plant Pathology and Plant Protection, University Göttingen, for some literature searches and for assistance in delivering, receiving and converting emails, and the efforts by Mr. W. Eisenberg, M.A., Institute of

Subtropical Agriculture, University Göttingen, for help in preparing figures on the PC.

H.P. is especially indebted to his wife for her many helpful and elucidating discussions, and for her patience and encouragement during the years when writing the book.

Göttingen and New Brunswick,   Hermann H. Prell
December 2000   Peter R. Day

Contents

List of Tables

List of Figures

1 Introduction

The interactions between plants and phytopathogenic fungi are complex. The first studies of the processes involved pursued two questions: first what is the physiological and biochemical nature of the interaction; and second how are the interactions genetically determined? The latter approach was particularly important for breeding crops resistant to attack by phytopathogenic fungi. Different experimental systems were employed depending on the view-point and the questions emerging from it. Numerous observations and experiments with many different host-pathogen systems led to the discovery of a great many singular causal connections. Assembling these facts to make a rational and homogeneous picture of the processes going on during plant-pathogen interactions required a series of working hypotheses, some of which still have to be proved or disproved.

During the last twenty years it has become increasingly clear that substantial progress in understanding plant-pathogen interactions can only be achieved if the action of the genes involved can be characterized as precisely as possible, not only formally in terms like "suppression", "activation" or "recognition". From this type of analysis, plant pathologists were able to incorporate experimental results into working hypotheses that were derived from genetic analyses and physiological and biochemical investigations. This has led to the formulation of many new questions, which could be tackled by molecular biology and molecular genetics with newly designed experimental systems. The same new tools were also applied to reconsider the older working hypotheses.

Although we are still far from having answers to all of the open questions, our present knowledge and derived hypotheses offer a basis for attempting to integrate systematically many observations and conclusions. This book describes the complex network of plant-pathogen interactions with a general picture in mind. Some especially interesting new findings, and the conclusions and hypotheses emerging from them, will be described in more detail. Well established experimental systems, and their genetic backgrounds, are emphasized in the hope of inspiring the reader to ask further questions and/or to think about novel experimental systems.

To help the reader facts or ideas presented in other chapters are often repeated. Cross references are also given to chapter(s), where a topic is discussed in more detail or in a different context. The glossary and index are a further guide to facts and definitions. Suggestions for further reading are given at the end of each chapter and are divided into "reviews" and "relevant papers". Only "relevant papers" are cited in the text and refer to particular experimental work. The reviews cited are of a more general nature and are generally not cited within the text.

The following synopsis was prepared for readers who prefer an overview before becoming immersed in details. It contains commonly used technical terms (boldface printed in this chapter) with short, clear definitions and examples of their use. Unequivocal communication or description of phenomena and their connections requires clearly defined terms. Readers who find this dry and abstract can skip this section and start with chapter 2. In all the chapters new terms are defined and explained where they are first used. Should the reader get lost he can go back to the overview for re-orientation. Definitions of technical terms can also be obtained via the index. Page numbers in italics refer to the glossary.

How do plants and phytopathogenic fungi behave if they meet each other? Ordinarily, plants reject attacking phytopathogenic fungi. Parasitism is an exception to the rule. The result of the interaction between both partners, parasitic colonization of the plant by the pathogen, or failure of parasitism because the pathogen is repelled by the plant's defenses, depends first of all on the genetic determinants of the plant and the pathogen. Most plants are not normally colonized and parasitized by most pathogens. They are **non-host plants** for these pathogens. Non-host plants were sometimes described as exhibiting **immunity** against pathogen attack. The reasons why plants are non-hosts may differ in detail, but in general the term designates a situation where the pathogen is unable to surmount barriers that prevent colonization of the plant. These various barriers comprise the **basic resistance** or **basic incompatibility** of the plant. The underlying mechanism of basic incompatibility may depend either on the plant, on the pathogenic fungus, or on both. Plant mechanisms may also protect against attack by quite different plant pathogens such as arthropods, nematodes, phytopathogenic bacteria or viruses. Occasionally the term **avoidance** is used when a plant manages to resist a pathogen attack, irrespective of the mechanism involved.

As mentioned above, successful attack of a plant by a pathogenic fungus is a rare exception and not the rule. Successful parasitism by the fungus depends on the production of **pathogenicity factors** acquired during evolution. As result of successful attack by a pathogenic fungus the plant becomes a **host plant** for that pathogen. Between both partners there exists a so called **basic compatibility**. This term embodies all physiological and biochemical requirements for colonizing and parasitizing a particular plant by a particular pathogenic fungus. During the process of colonization the pathogen withdraws nutrients from its host plant and lives and multiplies at its expense. This leads to biochemical changes within the plant that result in more or less conspicuous **disease symptoms** like yellowing of leaves, wilting, necrosis or distortions of plant form which reduce the vitality of the plant and, with time, may finally end in the death of plant parts or even the whole plant. – Basic compatibility is a highly specific phenomenon referring to only a particular plant species and the appropriate pathogen species or *forma specialis*. On the contrary, basic resistance and basic incompatibility are rather unspecific phenomena observed among plants in contact with almost all pathogens.

Even though plant and pathogen may have evolved basic compatibility, the host plant may deploy defense strategies to limit pathogen attack. Two basically different defense strategies can be distinguished in terms of the physiological and biochemical mechanisms that result in resistance:

(a) The plant cell(s) injured by the invading pathogen dies very rapidly causing necrosis of immediately adjacent tissue. In this way the pathogen is cut off from its nutritional substrate, is thus prevented from attacking further plant cells, and finally dies by starvation. This is the so called **hypersensitive reaction (HR)** or **hypersensitive cell death.** It is a rapid, and in many cases, macroscopically observable necrotic response of plant cells to pathogen attack.

(b) The plant cell survives the pathogen attack and is induced to build up new defense barriers by *de novo* synthesis of mRNA and protein. This is called a **non-hypersensitive reaction.** HR can also lead to this response in adjacent tissues.

The interaction of a **host plant** with a basically compatible pathogen is called a **homologous interaction.** The pathogen able to parasitize this plant is an **homologous pathogen.** The interaction of a **non-host plant** with a pathogen is designated a **heterologous interaction.** In this case the pathogen is a **heterologous pathogen,** the plant exhibits basic resistance, and no colonization can ensue. However, when a pathogen interacts with certain non-host plants, a hypersensitive reaction may be triggered which then is termed a **heterologous hypersensitive reaction** or **heterologous HR.**

The homologous interaction of a pathogen with its host plant, and its colonization, is also called a **compatible** interaction of both partners. However, under certain conditions, such as the host plant has a gene for resistance against a particular homologous pathogen, the interaction becomes **incompatible.** This means the pathogen is rejected by its host plant although there is basic compatibility between both partners. This so called **host resistance** comes into action because the infecting pathogen induces defense reactions in the host plant. These block or restrain growth and reproduction of the parasite thereby restricting colonization of the plant. Host resistance presupposes, at least by definition, the existence of basic compatibility between the partners.

In the plant pathology literature, the terms compatible and incompatible are usually applied only to homologous interactions, i.e. to the interactions of a host plant with homologous pathogens. **Homologous compatibility** leads to colonization of the host plant and survival of the pathogen. On the other hand, **homologous *in*compatibility** means that the pathogen can neither parasitize its host plant nor survive. An exception to this is the so called **heterologous *in*compatibility** which designates a **heterologous HR**, i.e. the appearance of an HR reaction between pathogens and certain non-host plants. – In cases where a compatible interaction does not result in the appearance of disease symptoms on the plant, or a significant reduction of crop yield, one speaks of **tolerance** of the plant towards the pathogen.

Phytopathogenic fungi follow two different strategies to ensure their nutrition; necrotrophy and biotrophy. **Necrotrophs** first kill their host cells before they can colonize them. Killing is brought about either by changes in the cell metabolism after infection, or is due to the action of **toxins** or **extracellular enzymes** produced by the infecting pathogen. (Necrotrophy of fungal pathogens is sometimes also called **perthotrophy** thus emphasizing these pathogens first kill their host cells before parasitizing them. In that case the term necrotrophy is confined to saprophytic, non-pathogenic fungi unable to kill plant cells and only feeding on already dead plant materials.) According to the range of activities of toxins synthesized by plant pathogens one distinguishs between **host specific toxins** or **host-selective toxins** on one side and **non-host-specific toxins** or **non-host-selective toxins** on the other.

Biotrophs depend on the metabolism of the infected host cells and surrounding plant tissue. This means that injury leading to host and tissue death is usually delayed until reproduction of the pathogen has been completed. Some biotrophs display during their life cycle first a biotrophic phase and later on, when the plant tissue collapses and dies, a saprophytic phase that exploits the metabolites present in the dead tissue.

As mentioned already, phytopathogenic fungi attack only particular plant species, i.e. they exhibit a limited **host range**. This means a pathogen can overcome the basic resistance of a plant within its host range and establish with it a basic compatibility. The host range of a pathogen is determined by the activities of its **pathogenicity genes** and their **pathogenicity factors**, respectively, "tuned" for this very host plant. In other words, distinct pathogenicity factors determine the specificity of the pathogen for a particular plant. However, pathogens such as *Botrytis* or several *Phytophthora* spp. exhibit little or no specificity for distinct host plants. Some pathogen species can be subdivided into *formae speciales* (**f.sp.**), comparable to the subspecies of a plant. The members of a *forma specialis* are generally confined to a single plant species (**species specificity**), or more rarely to several species.

The different defense mechanisms employed by plants against attacking pathogens can also be distinguished by the level at which they come into action: A pathogen is first faced with basic resistance (non-host resistance, general resistance, basic incompatibility, broad resistance or parasite non-specific resistance) which represents the **first level** of **pathogen defense** in a plant. When basic resistance is effective, the pathogenic fungus succeeds at best in establishing a transient contact with the plant, but is unable to colonize it. As mentioned above, this is valid for most pathogens, except those for which the plant serves as a host. However, if this is the case, the pathogen actively breaks the defense barriers of basic resistance by employing its pathogenicity genes. These barriers may have existed before infection or have been induced to develop in the plant by the infecting pathogen. However, changes in the plant's physiological state may interfere with or hinder the expression of basic resistance.

Even when basic compatibility has been established and the plant is a host for the pathogen, other barriers may prevent parasitism and colonization. As mentioned above, a host plant may be resistant to attack by a particular homologous pathogen by inheriting a gene for resistance. In this case, the plant can present a **second level** of **pathogen defense** which represents **host resistance**. These defense mechanisms are, in contrast to basic resistance, highly specific for the pathogen since only certain races of a pathogen species will be rejected by the resistant cultivar. For that reason, this kind of defense is designated **cultivar-specific resistance** or **parasite specific resistance**. Other homologous pathogen races may not be affected by the particular cultivar-specific resistance and therefore are still able to parasitize that plant. Cultivar specific resistance is almost exclusively observed against attack by biotrophic pathogens. – In summary, the characteristic difference between basic resistance and host resistance of plants consists in the specificity of pathogen defense: The former is unselectively directed against all attacking pathogens, the latter is highly selective against only one particular pathogen species.

There are two different types of host resistance genes: **race-non-specific resistance** (horizontal resistance, uniform resistance or generalized resistance) and **race-specific resistance** (vertical resistance, specific resistance, or differential resistance). Race-non-specific (horizontal) resistance is directed against all members or races of a pathogen species. In contrast, race-specific (vertical) resistance confers resistance only against particular pathogen races. In the latter case a particular mutation to race-specific resistance in the host plant can specifically "recognize", i.e. discriminate between, different races of the pathogen. Genetic determinants active against one particular pathogen race might even be found in different plant genera. The mechanisms and genetic determination of horizontal and vertical resistance are quite different. Both defense mechanisms are apparent only in the presence of basic compatibility.

Resistance, or the prevention of plant parasitism, can result from two situations: either the plant is not a host for the pathogen, i.e. the pathogen is blocked by basic resistance or basic incompatibility, or the plant is indeed a host but carries genes for parasite- or cultivar-specific resistance, which enable it, in spite of existing basic compatibility, to reject the pathogen on a second level of defense. Both resistance responses are similar and in some cases it may be difficult to determine whether defense is due to basic or cultivar-specific resistance. The specificity of plant resistance to pathogens can be summarized as follows: basic resistance (shown by non-host plants) is directed against nearly all pathogens, race-non-specific resistance (shown by host plants) is against all races of a pathogen, or of a *forma specialis*, and finally race-specific resistance (also shown by host plants) is only against certain races of one particular pathogen or *forma specialis*.

In other words basic resistance (non-host resistance) and cultivar specific resistance (host resistance) are both terms that describe the inability of a phytopathogenic fungus to parasitize a plant. However, this inability refers to different situations: in cultivar-specific resistance (both horizontal and vertical) we may suppose the wild type host plant acquired by mutation(s) the ability to prevent a particular homologous pathogen from colonizing it. On the other hand, the term basic resistance designates the situation in which a wild type non-host plant can not be colonized by heterologous pathogens because the plant, the pathogen, or both, lack appropriate genes. Since the term "resistance" in colloquial usage refers only to a property of the plant but not of the attacking pathogen it would be more accurate to use the more comprehensive term **basic incompatibility** instead of basic resistance. In the phytopathological literature however, basic resistance is the more widely used term. – Recent research suggests that the molecular basis of release mechanisms for active non-host resistance (or basic incompatibility) and for race-specific or vertical resistance are quite similar or even alike.

Quite different defense mechanisms are observed against necrotrophic phytopathogenic fungi that produce host-specific toxins. Mutation to resistance against a host-specific toxin entails either its detoxification by the host plant or that the plant becomes insensitive to its action.

The colonization of a plant by a pathogenic fungus can be summarized as follows: First the pathogen can parasitize a plant only after it has overcome the plant's basic resistance and second, if it encounters host resistance, the pathogen must negate the corresponding race-specific resistance determinant. Overcoming race-non-specific resistance by pathogens has so far not been observed, whereas negating and overcoming race-specific resistance is by no means rare, since generally only one mutation of the pathogen to **specific virulence** is sufficient.

The term virulence is generally used in a quantitative sense to describe the extent of host plant symptoms. In contrast, specific virulence of a biotrophic pathogen describes a qualitative character, i.e. the ability, acquired as a result of a single step mutation, to overcome the race-specific resistance of the host plant. Specific virulence is either present or absent. Corresponding to the host range determined by pathogenicity genes one can define a "second" type of host range delimited by the race-specific resistance properties of the cultivar. Therefore, pathogens may have two different specificities: their **species specificity**, dependent on their possession of appropriate pathogenicity genes, and their **race-specificity** dependent on their ability to overcome race-specific resistance determinants in the host plant by mutation of the pathogen to specific virulence.

Reviews

Bailey J.A. (1983): Biological perspectives of host-pathogen interactions. In: Bailey J.A., Deverall B.J. (eds.): The Dynamics of Host Defence. Academic Press Australia, Sidney, New York, London, Paris, San Diego, San Francisco, Sao Paulo, Tokyo, Toronto. 1 – 32
Day P.R. (1974): Genetics of Host-Parasite Interaction. W.H.Freeman & Co., San Francisco
Ellingboe A.H. (1976): Genetics of host-parasite interactions. In: Heitefuss R., Williams P.H. (eds.): Encyclopedia of Plant Physiology, (NS), Physiological Plant Pathology. Springer Verlag, Heidelberg. 761 – 778
Heath M.C. (1991): Evolution of resistance to fungal parasitism in natural ecosystems. New Phytol. 119: 331 – 343
Knogge W. (1991): Plant resistance genes for fungal pathogens – physiological models and identification in cereal crops. Z.Naturforsch. 46c: 969 – 981
Shaner G., Stromberg E.L., Lacy G.H., Barker K.R., Pirone T.P. (1992): Nomenclature and concepts of pathogenicity and virulence. Annu.Rev.Phytopathol. 30: 47 – 66
Sprecher E., Urbasch I. (1984): Wechselwirkungen zwischen Pflanzen und pathogenen Pilzen. Naturwiss.Rundsch. 37: 401 – 407

2 Recognition, the First Step in Interaction Between Plant and Pathogen

Interactions between plants and pathogenic fungi commence with some kind of "communication" between them. If during this contact some specific requirements are met, the so called "recognition" between both partners can take place. Recognition implies that the onset of a series of biochemical reactions is triggered in one or both partners. The recognition event consists of the specific interaction of a particular signal with its sensor or receptor, the latter being located on the surface of one of the interacting partners. This kind of interaction is called a signal-sensor reaction. The signal can be either chemical or physical (e.g. thigmotropic signal) in nature such as some morphological attribute of one of the partners. Signal recognition requires very close contact between plant and pathogen. For chemical signals this condition is met for example during penetration of the host cell wall by the pathogen. During this process the pathogen itself comes in close contact with the plant cell membrane which carries suitable sensor or receptor sites for recognition. Thigmotropic signals may be recognized by pathogen germ tubes sensing the precise location (Allen et al. 1991) of the vascular bundles in the leaves of their host plants or the shape of their stomata and the openings between their guard cells. The sensors are located in the tip of the germ tube guiding the pathogen into the intercellular spaces of the leaves.

The direct recognition of a chemical signal by its sensor may be regarded as the interaction between a ligand and its receptor, a process that depends on their steric properties. Binding is assumed to trigger further reactions that may either promote or impede penetration and/or growth of the fungal hyphae. The molecular basis of the reception of thigmotropic signals is presently still not clear. Hoch and Staples (1991) proposed for example that thigmotropic sensing depends on a deformation of the plasma membrane of the fungal germ tube resulting in a local stress reaction effecting influx of Ca^{++} ions.

Recognition between plant and pathogen may comprise two levels of interaction. On the first level, the (homologous) pathogen recognizes the plant as a host and establishes basic compatibility since it can present to the plant "appropriate" pathogenicity genes for parasitizing it (see also chapter 7). However, if the host plant also carries a gene for race-specific resistance directed against the infecting pathogen (see chapter 9.1), a second level of recognition occurs and the plant now recognizes the presence of the attacking "corresponding" pathogen. This host recognition triggers the expression of defense reactions against the invader. The phenomenon of race-specific resistance was thoroughly analyzed by Flor, the father of the genetics of race specific resistance. In his early papers he called a very strong expression of host plant resistance "immunity" (Flor 1942). At a later date the term immunity was used to describe the lack of interaction between non-host plants and heterologous pathogens as mentioned in chapt. 1.

Recognition events that promote the establishment of basic compatibility trigger in the pathogen the expression of its pathogenicity genes. For example the phytoalexin pisatin of pea plants induces in the infecting pathogen *Nectria haematococca* synthesis of demethylase, an enzyme that detoxifies pisatin. In other words, the pisatin synthesized by the host is recognized by *N. haematococca* as a signal for the synthesis of a pathogenicity factor which neutralizes the plant defense reaction. A pathogen can also induce metabolic changes in its host plant which prevent the expression of basic resistance. By this means the pathogen induces the host to support its development. Once basic compatibility has been established an infecting pathogen may also be induced to synthesize an avirulence factor. For example the product of the avirulence gene *Avr9* of *Cladosporium fulvum* is only produced by the pathogen living inside a tomato plant, not in axenic culture (except that this was done under conditions of nitrogen starvation). Apparently, synthesis of the avirulence factor *in planta* depends on a signal of plant origin.

The reactions that occur during the recognition of pathogens by plants have been investigated in different experimental systems employing different elicitors, HR-elicitors triggering an hypersensitive reaction (HR) on race-specific resistant host plants, or non-HR elicitors triggering a non-hypersensitive reaction like phytoalexin synthesis on non-host plants. However, the most informative biochemistry emerged from investigations where the host and parasite genes have been identified (see also chapter 9.1.4). At this place only some salient observations shall be mentioned relevant to both systems. The first reactions, detectable a few minutes after the recognition event triggered by an invading pathogen, are cytological alterations inside the affected cell such as movement of the cell nucleus and the cytoplasm towards the penetration site. The same movements of cell components are also observed within tissue in cells neighbouring the cell penetrated by the pathogen. Furthermore in both systems the cell membrane was shown to be depolarized and its permeability increased. The most prominent change is the so called K^+/H^+ response resulting in leakage of K^+ ions out of, and movement of H^+ ions into, the plant cells (chapter 9.1.3.3). Moreover a so called oxidative burst (Apostol et al. 1989; Orlandi et al. 1992) is observed in which active oxygen is produced in the form of superoxide anions (O_2^-), perhydroxyl radicals (HO_2), OH radicals, and hydrogen peroxide (H_2O_2). The oxidative burst results in cross-linking of proline rich structural proteins within the plant cell wall. All of these reactions appear to be obligatory intermediate steps before transcription activation takes place for the expression of plant defense reactions such as the synthesis of phytoalexins, or the deposition of lignin in cell walls. The synthesis of hydrolytic enzymes, which degrade the cell walls of the parasitizing fungus, is also induced. All of these compounds employed in defense against attacking pathogens are termed defense compounds.

Whether a defense reaction can be attributed to active basic resistance (directed against a very broad spectrum of pathogen species) or to race-specific resistance (directed only against single pathogen races) depends on the strategy employed by the fungal pathogen to parasitize its host, i.e. by necrotrophy or by biotrophy (see chapter 3). Regardless of its strategy, the pathogen is first confronted with the passive and/or active basic resistance of the plant which is effective against nearly all plant pathogens.

When a necrotrophic pathogen is recognized and repulsed by a plant, the distinction between active basic resistance and race-specific resistance is usually meaningless because necrotrophs generally kill their host cells immediately after infection. Thus even host plants carrying a corresponding gene for host resistance would have no chance to build new defense barriers against the pathogen. However, in the tissue surrounding the killed cell the synthesis of new defense barriers could be triggered by signal substances that spread out from it. Many necrotrophs show little species specificity in overcoming the basic resistance of plants, i.e. they exhibit a broad host range. However, there are some exceptions: for example the genus *Helminthosporium* shows a high specificity only infecting oat, maize, or sugar-cane, *Colletotrichum lindemuthianum* only French beans, *Phoma lingam* only rape and *Septoria nodorum* only wheat.

Unlike necrotrophs, biotrophic pathogens generally have a much narrower host range and do not kill their host plants after infection. Rather they leave the infected cell alive as long as possible because they depend on the unimpaired metabolism of their host cells for their nutrition and reproduction. This situation offers a plant with genes for host resistance the opportunity to raise new defense mechanisms against the pathogen even though its basic resistance has been overcome and basic compatibility established. This second level of defense is a form of host plant resistance. This so called vertical resistance is directed solely against particular races (see chapters 4 and 9.1) of one pathogen species. Plants carrying a gene encoding this highly specific recognition function probably emerged by mutation which occurred independently of the presence of the pathogen. Race-specific resistance is highly selective and only effective against particular pathogen races. This is in strong contrast to the unspecific recognition of many different pathogen species as observed in active basic resistance. (The recognition processes of race-specific resistance will be discussed in detail in chapters 9.1.3.1 and 9.1.4). The molecular basis of recognition in active basic resistance may be very similar or even identical to that of race-specific resistance.

In this context it is important to stress that there is still another type of cultivar-specific resistance, the so called race-non-specific resistance or horizontal resistance. In contrast to race-specific resistance, this is effective against all races of one *forma specialis* of a pathogen. To date there is no experimental evidence for recognition events occurring between pathogen and host plant for the expression of race-non-specific resistance (for details see chapter 9.2). The genetic determination of race-non-specific resistance is quite different from that of race-specific resistance.

The recognition events between pathogens and plants that lead to expression of defense reactions have been attributed to signal-sensor reactions consisting of three different components: (1) the elicitor, a low molecular weight substance which is the signal originating from, or depending on the action of, the pathogen. (2) the sensor or receptor of the plant which specifically binds the elicitor, and (3) the effector, which designates one or more substances formed or activated as a result of the recognition and binding between elicitor and receptor. Effectors comprise all the products participating in signal transduction for triggering expression of defense reactions by the plant. The process of elicitor-receptor-interaction is called elicitation. In this context the elicitor acts like a ligand binding to the receptor. Both, elicitation and the formation of effectors release the expression of both, active basic resistance and race-specific resistance.

Since recognition between plants and their pathogens is based on signal-sensor interactions, an obvious question is: what kind of substances can function as chemical signals? Are any substances candidates for elicitation or only those belonging to particular classes? To investigate this question it is helpful to differentiate between the chemicals that trigger active basic resistance and those that trigger race-specific resistance. Triggering active basic resistance would appear to be rather unspecific with respect to the pathogen (but not unspecific with respect to the chemical and steric nature of both the elicitor and the receptor constituting the base of the signal-sensor interaction!), whereas triggering of race-specific resistance exhibits a very high specificity even with respect to the pathogen race concerned.

Expression of active basic resistance is triggered if the plant recognizes any event that impairs or destroys its integrity. This happens for example when plant cell walls are injured by pathogen exoenzymes. These may split chemical bonds within the plant cell wall releasing breakdown products that elicit defense reactions. Since these elicitors originate from the plant they are called endogenous elicitors of active basic resistance. They may be monomers or oligomers of the cell wall polymers. However, endogenous elicitors may also result from mechanical wounding and the release of plant enzymes that break down the plant cell wall.

Active basic resistance can also be induced by elicitors that originate from the attacking pathogen. These exogenous elicitors may be inherent pathogen surface molecules or low molecular weight breakdown products, such as glucan and chitosan, released from the pathogen cell wall by the action of chitin degrading enzymes produced as defense compounds by the plant. These elicitors may be classified according to their origin, either the cell walls of plants or pathogens. The absence in the plant of specific receptors for recognizing endogenous or exogenous elicitors results in the establishment of basic compatibility between both partners. This assumes that the pathogen has all the necessary pathogenicity genes to parasite its host plant.

Unlike the establishment of basic resistance, race-specific resistance requires that the plant has a gene for resistance which specifically recognizes only particular races of the pathogen. As in active basic resistance the specific receptors engaged in race-specific resistance recognize only distinct elicitors, the so called corresponding avirulence factors, which are synthesized by the pathogen. Because of their high specificity, avirulence factors are also designated specific elicitors. All avirulence factors of phytopathogenic fungi so far identified by molecular genetic methods are polypeptides or proteins.

Endogenous elicitors and exogenous elicitors are low molecular signal substances specifically recognized by plant receptors that trigger the expression of plant defense reactions. Their designation as endogenous or exogenous refers to whether they are derived from the plant or the pathogen. In the earlier literature, enzymes which liberate substances from the plant cell wall that effect elicitation were also called elicitors. However, this does not fit the definition of an elicitor as a low molecular weight signal substance. In all cases where enzymes were called elicitors low molecular weight signal molecules were subsequently identified. Only most recently it was discovered that even a high molecular weight protein, xylanase, may act as an elicitor under conditions that excluded elicitation being effected by digestion products of the enzyme substrate (Sharon et al. 1993).

An attacking pathogen can override or bypass the active basic resistance of a plant only if the plant recognition system fails. For signal-sensor reactions this could happen in four different ways: (1) Binding between elicitor and receptor could be prevented by a mutation causing either the absence of or a structural change in the elicitor, (2) the receptor could be masked by another substance, (3) the elicitor could be competed off the receptor by a closely related substance – formally a suppression of recognition –, or (4) the receptor could be changed by mutation to prevent recognition of the elicitor. In each case the pathogen can parasitize the plant if it is endowed with the "appropriate" pathogenicity genes. – Similar mechanisms, involving different elicitors and receptors, might also prevent expression of race-specific resistance.

Plant recognition of elicitors originating either from the plant or the pathogen could occur together by the same, or very similar, recognition systems. These may recognize both "self-produced" signals, which induce basic resistance, and "foreign" signals which induce race specific resistance. In fact active basic resistance and race-specific resistance are triggered by different signals. For example active basic resistance is triggered by particular oligosaccharides which are ineffective in triggering race specific resistance. Whether, and to what extent, the participating receptors are equal to each other, irrespective of their different specificities for elicitors and the accompanying processes for signal transduction, is so far not known.

Signal-sensor reactions and the accompanying signal transduction processes will be discussed in detail in sections 9.1.3.1, 9.1.3.3 and 9.1.4.

Reviews

Atkinson M.M. (1993): Molecular mechanisms of pathogen recognition by plants. Advances in Plant Pathology **10**: 35 – 64

Bailey J.A. (1983): Biological perspectives of host-pathogen interactions. In: Bailey J.A., Deverall B.J. (eds.): The Dynamics of Host Defence. Academic Press Australia, Sidney, New York, London, Paris, San Diego, San Francisco, Sao Paulo, Tokyo, Toronto. 1 – 32

Boller T (1995): Chemoreception of microbial signals in plant cells. Annu.Rev.Plant Physiol.Plant Mol.Biol. **46**: 189 – 214

Daly J.M. (1984): The role of recognition in plant disease. Annu.Rev.Phytopathol. **22**: 273 – 307

Dixon R.A., Harrison M.J., Lamb C.J. (1994): Early events in the activation of plant defense responses. Annu.Rev.Phytopathol. **32**: 479 – 501

Doke N., Miura Y., Sanchez L.M., Yoshioka H., Kawakita K. (1994): Signal transduction and hypersensitive reaction in *Phytophthora*-Solanaceae plant pathosystem: superoxide generating reaction. In: Kohomoto K., Yoder O.C. (eds.): Host Specific Toxin: Biosynthesis, Receptor and Molecular Biology. Faculty of Agriculture, Tottori University, Sogo Printing and Publishing Co., Ltd., Tottori, Japan. 153 – 167

Ebel J., Cosio E.G. (1994): Elicitors of plant defense responses. Int.Rev.Cytology **148**: 1 – 36

Halverson L.J., Stacy G. (1986): Signal exchange in plant-microbe interactions. Microbiol. Rev. **50**: 193 – 225

Hammond-Kosack K.E., Jones J.D.G. (1996): Resistance Gene-Dependent Plant Defense Responses. Plant Cell **8**: 1773 – 1791

Heath M.C. (1991): Evolution of resistance to fungal parasitism in natural ecosystems. New Phytol. **119**: 331 – 343

Heath M. (1994): Current concepts of the determinants of plant-fungal specificity. In: Kohomoto K., Yoder O.C. (eds.): Host Specific Toxin: Biosynthesis, Receptor and Molecular Biology. Faculty of Agriculture, Tottori University, Sogo Printing and Publishing Co., Ltd., Tattori, Japan. 3 – 21

Hoch H.C., Staples R.C. (1991): Signaling for infection structure formation in fungi. In: Cole G.T., Hoch H.C. (eds.): The Fungal Spore and Disease Initiation in Plants and Animals. Plenum Press, New York and London. 25 – 46

Keen N.T. (1982): Mechanisms conferring specific recognition in gene-for-gene plant parasite systems. In: Wood R.K.S. (ed.): Active Defence Mechanisms in Plants. Plenum Press, New York, London. 67 – 84

Keen N.T. (1982): Specific recognition in gene-for-gene host-parasite systems. Advances in Plant Pathology **1**: 35 – 81

Knogge W. (1991): Plant resistance genes for fungal pathogens – physiological models and identification in cereal crops. Z.Naturforsch. **46c**: 969 – 981

Knogge W. (1998): Fungal pathogenicity. Curr. Opinion Plant Biol. **1**: 324 – 328

Kolattukudy P.E., Podila G.K., Sherf B.A., Bajar M.A., Mohan R. (1991): Mutual triggering of gene expression in plant-fungus interactions. In: Hennecke H., Verma D.P.S. (eds.): Advances in Molecular Genetics of Plant-Microbe Interactions. Dordrecht, Kluwer Academic Publishers, Netherlands. 242 – 249

Kolattukudy P.E., Rogers L.M., Li D.X., Hwang C.S., Flaishman M.A. (1995): Surface signaling in pathogenesis. Proc.Natl.Acad.Sci.USA **92**: 4080 – 4087

Manocha M.S., Sahai A.S. (1993): Mechanisms of recognition in necrotrophic and biotrophic mycoparasites. Can.J.Microbiol. **39**: 269 – 275

Ryan C.A., Farmer E.E. (1991): Oligosaccharide signals in plants: a current assessment. Annu.Rev.Plant Physiol. **42**: 651 – 674

Scheel D., Parker J.E. (1990): Elicitor recognition and signal transduction in plant defense gene activation. Z.Naturforsch. **45c**: 569 – 575

Sutherland M.W. (1991): The generation of oxygen radicals during host responses to infection. Physiol.Mol.Plant Path. **39**: 79 – 94

Tzeng D.D., DeVay J.E. (1993): Role of oxygen radicals in plant disease development. Advances in Plant Pathology **10**: 1 – 34

Relevant papers

Allen E.A., Hazen B.E., Hoch H.C., Kwon Y., Leinhos G.M.E., Staples R.C., Stumpf M.A., Terhune B.T. (1991): Appressorium formation in response to topographical signals by 27 rust species. Phytopathology **81**: 323 – 331

Apostol I., Heinstein P.F., Low P.S. (1989): Rapid stimulation of an oxidative burst during elicitation of cultured plant cells: role in defense and signal transduction. Plant Physiol. **90**: 109 – 116

Flor H.H. (1942): Inheritance of Pathogenicity in *Melampsora lini*. Phytopathology **32**: 653 – 669

Orlandi E.W., Hutcheson S.W., Bakers C.J. (1992): Early physiological responses associated with race specific recognition in soybean leaf tissue and cell suspensions treated with *Pseudomonas syringae* pv. *glycinea*. Physiol.Mol.Plant Path. **40**: 173 – 180

Sharon A., Fuchs Y., Anderson J.D. (1993): Elicitation of ethylene biosynthesis by a Trichoderma xylanase is not related to the cell wall degradation activity of the enzyme. Plant Physiol. **102**: 1325 – 1329

3 Activities of Phytopathogenic Fungi when Colonizing Their Host Plants

To colonize and parasitize a plant the pathogen has to solve two problems: first it must reach the interior of the plant and second it has to establish itself there to be able to exploit the host plant. The first step requires the pathogen to overcome the defense barriers of basic resistance in order to establish basic compatibility. For example this may require the germ tube of a spore to express particular fungal genes that allow it to penetrate the host plant cell wall. The subsequent colonization of the plant also depends on the expression of fungal genes which, in the absence of competing microorganisms, make the new living space accessible to the parasite. The fungal genes required for parasitizing a plant are called pathogenicity genes. Their products, the pathogenicity factors, provide for the many different functions needed. Pathogenicity genes are distinct from the so called "house keeping" genes of the pathogen. The latter are required only for sustaining cell functions like those occurring during growth on artificial media. Consequently, a mutation to auxotrophy, which renders a pathogen dependent on an external supply of, for example, tryptophan for growth on an artificial medium or in a host plant, is not considered a mutation in a pathogenicity gene even though the auxotrophic pathogen can not parasitize its host plant in the absence of the supplement. Thus pathogenicity genes serve the pathogen only for invading and utilizing the niche represented by the "plant". They are not required for completing a whole life cycle in axenic culture – starting with spore germination and proceeding through mycelial growth and production of new progeny. However, for most obligate biotrophic pathogens we cannot easily distinguish between house keeping and pathogenicity genes, since these pathogens can only complete their life cycle in a living plant.

The pathogenicity genes involved in the interactions between fungal parasites and plants can be quite varied since they have to provide for many single steps such as penetration, formation of special morphological structures required for colonizing the plant, establishing the necrotrophic or biotrophic modes of nutrition, and so on. Within these long reaction chains a single gene activity may be involved in one or more of the empirically defined steps of interaction. Thus a single gene may encode both a biochemical function and the formation of a matching morphological structure. Since the outcome of plant-pathogen interactions also depends on the properties and activities of the plant, the expression of pathogenicity genes may even result in the defeat of the attacking pathogen.

This is the case with pathogenicity genes that encode plant cell wall digesting enzymes. Among the cell wall breakdown products there may be compounds that elicit the defense reactions of basic resistance. Thus, pathogenicity genes may ultimately prevent a pathogen from parasitizing a plant (see also chapters 2 and 6, on active basic resistance).

Spores or hyphae of pathogenic fungi may enter wounded plants, whereas saprophytic fungi cannot colonize living plants even when they can enter wounds. This characteristic difference in behaviour may be considered as the first step towards the evolution of phytopathogenicity. However, opportunistic pathogenicity is observed only against stressed or senescent plants whose basic resistance is already reduced. Such pathogens exhibit little host specificity.

Spores of most phytopathogenic fungi are spread by wind. When they land by chance on a leaf of their host plant they are fixed to its hydrophobic surface by a secreted mucilaginous droplet, or by liberating enzymes that digest and soften the substrate. Spores of parasitic fungi can, after sufficient hydration, germinate in absence of nutrients. Invasion by germ tubes or hyphae may proceed through natural openings like stomata or lenticels, which are actively sought out by the growing hyphae. In other cases the hyphae penetrate the epidermis immediately after spore germination by secreting cutinases as pathogenicity factors that digest the cuticle, and cellulases, pectinases and pectinesterases that digest the cell wall structures lying underneath. Penetration of epidermal cells, and cells accessible from the intercellular spaces, is supported by appressoria which attach firmly to the cell wall. In some cases the appressorium contains a penetration peg which mechanically perforates the plant cell wall and in others penetration involves both enzymatic and mechanical perforation. Generation of the mechanical pressure for piercing the cell wall by the peg inside the appressorium seems to depend on the synthesis of melanin. In *Colletotrichum lagenarium* and *Magnaporthe grisea* the melanin forms inside the appressorium within a semipermeable membrane which, by enabling the build-up of hydrostatic pressure, allows the peg to pierce the cell wall (Kubo et al. 1991, Kimura & Tsuge 1993). An albino mutant of *C. lagenarium* forming colorless appressoria was nonpathogenic. Restitution of the defective melanin synthesis by transformation with DNA derived from a wild type gene bank restored pathogenicity.

The pathogenicity factors play an important role in overcoming basic resistance. For example enzymatic digestion enables hyphae to penetrate even a thick epidermis. Other pathogenicity factors are directed against physiological plant defense mechanisms such as the detoxification of the phytoalexin pisatin synthesized in pea plants, *Pisum sativum*, in response to attack by the necrotroph *Nectria haematococca*. In this case the pathogenicity factor synthesized by *N. haematococca* is a monooxygenase which demethylates the pisatin abolishing its function in pathogen defense. Generally a parasite's pathogenicity genes do not turn off the expression of host cell defense genes, rather they enable the pathogen to negate them. The attacked plant cell continues to express its defense genes even if basic compatibility is already established by the fungal pathogenicity factors.

The establishment of either necrotrophic or biotrophic nutrition of the pathogen provides further examples of the actions of pathogenicity genes. However, knowledge of the different gene activities involved are still lacking, particularly for biotrophic pathogens. Some pathogens employ both feeding habits. First a biotrophic phase is established which later on becomes necrotrophic after the break down of the plant tissue caused by the stress of biotrophic parasitism. Examples include pathogens of the genera *Colletotrichum*, *Phytophthora* and *Venturia*. Such parasites are called hemibiotrophic or facultative biotrophic pathogens.

Most necrotrophic pathogens first kill their host cells by excreting toxins, exoenzymes, or substances that act in a similar way to toxins such as polygalacturonidase or pectate lyase. Only after killing the host plant are necrotrophic pathogens able to colonize their host cells. Like saprophytes they decompose the plant materials by employing hydrolytic and proteolytic enzymes. The products of digestion are absorbed by the parasite.

The evolution of potent toxins with no host specificity, was a decisive step towards increasing phytopathogenicity. Necrotrophic pathogens that produce phytotoxins can still exist as saprophytes, but are inferior to true saprophytic fungi on dead plant material. However, necrotrophic fungi occupy a niche that is inaccessible to true saprophytes, since they can attack and kill living cells.

Some species of *Alternaria*, or of *Cochliobolus*, or its anamorph *Helminthosporium*, (an anamorph is a variant of a fungus unable to pass through a sexual cycle) synthesize host-specific toxins. These kill only certain cultivars which are sensitive to the toxin. Other cultivars of the same species may be insensitive. Toxin insensitive cultivars are still parasitized but do not exhibit disease symptoms. In several examples it has been shown that toxin sensitive plants have a specific receptor for the recognition of the host-specific toxin. (In chapters 8, 8.1, 8.2, and 8.3 the toxin producing pathogens will be treated in more detail.)

As already can be inferred from this short survey the killing of cells or whole plants by pathogen enzymes or toxins does not depend only on particular fungal gene(s) synthesizing an enzyme or toxin. A saprophyte is not converted into a necrotrophic pathogen by genetic transformation of a gene enabling it to synthesize a toxin. All necrotrophic fungi require the concerted action of many pathogenicity genes for the expression of pathogenicity.

In contrast to necrotrophs biotrophs establish themselves more "cautiously" within living and metabolizing host cells. Disease symptoms only appear relatively long after infection. Biotrophs exploit their host cells by forming haustoria. These are parts of their hyphae that project into the plant cells pushing in the host cell plasma membrane. They do not enter the plant protoplast like the true endoparasitic Chytridiomycetes and Plasmodiophoromycetes. They remain instead in the periplasmic space between the cell wall and the plasma membrane and from there withdraw nutrients from the protoplast. In many cases, biotrophs stimulate the metabolic activity of the infected cell and the surrounding tissue. The pathogen profits from this induced plant cell metabolism since its nourishment is enhanced. The areas of leaf tissue exhibiting such an increase in metabolism are called "green islands". Eventually, after progressive colonization of the host plant tissue, severe disease symptoms appear. These are the consequence of the stress provoked by the continual withdrawal of nutrients from the plant, and the accumulation of pathogen waste materials that finally results in its death. Thus, there is a close metabolic interaction between a plant and its biotrophic pathogen which is ultimately to the disadvantage of the plant. The high specificity involved in this interaction, as well as the specificity required for recognition between both partners, may be responsible for the generally limited host ranges observed for each *forma specialis* of a biotrophic pathogen.

The different strategies employed by pathogens for parasitizing their hosts are based on many different genes and physiological and biochemical activities. The most prominent characteristics of necrotrophic and biotrophic phytopathogenic fungi are shown in *Table 1*.

In the course of evolution encroachment of biotrophic pathogens into the metabolism of host cells probably became less pronounced, each step causing, at least in the beginning of infection, less stress to the plant and therefore less disruption of its metabolic activities. However, although the pathogen can utilize this food source for longer it becomes increasingly dependent on the host and finally may be unable to grow and reproduce outside it. This is true of obligate pathogens such as powdery mildew *Erysiphe* spp., the majority of rust fungi like *Puccinia* spp., *Uromyces* spp., potato wart *Synchytrium endobioticum*, and of clubroot *Plasmodiophora brassicae* none of which can be cultured in artificial media. So far the only exception seems to be *P. graminis* which today can be cultured to a limited extent on artificial medium.

In another form of specialization the infecting pathogen causes its host to synthesize a metabolite which only it can metabolize. For example the phytopathogenic bacterium *Agrobacterium tumefaciens* causes its host plant to produce certain types of aminoacids, the opines. The genetic information for their synthesis, i.e. the particular piece of DNA coding for the opine, is transferred

Table 1. Comparison of physiologic properties of necrotrophic or biotrophic phytopathogenic fungi and their host plants

	Necrotrophs	Biotrophs
Host range of the pathogen	Broad	Narrow
Axenic cultivation	Good	Poor
Production of toxins	Widespread	Seldom
Activity of secreted enzymes	High	Low

from the bacterium into the plant cell where it becomes integrated in the cell's genome. The bacterial gene directs the synthesis of the opine which serves the bacterium as a carbon and nitrogen source. In even more specialized cases the pathogen may also supply metabolite(s) to its host. If both partners benefit from their metabolic interdependence, the parasitic relationship is shifted to symbiosis. Examples include the bacterium *Rhizobium* which colonizes the root nodules of leguminosae, the lichens representing a symbiosis between fungi and algae, the mycorrhizas, found associated with the roots of many plants (Moore 1987), or fungal endophytes such as the recently discovered "sterile red fungus" (SRF, Lange 1992) isolated from wheat roots from Western Australia which has both disease suppressive effect (against *Gaeumannomyces graminis* var. *tritici*) as well as growth-promoting effect. The latter symbiotic relationship has two additional advantages for the plant: first it supports the growth of the plant and secondly it suppresses infection by pathogens since the SRF secretes a proteinaceous antibiotic. SRFs appear to be basidiomycetes since their mycelia have clamp connections.

Many of the pathogenicity genes of necrotrophs directly serve functions needed to overpower the host plant tissue. Their mode of producing toxins, may be compared to a racketeer who kills and exploits his victim. Once contacted the plant cell has no chance left to build up active defense mechanisms against the intruder but is killed more or less instantly. Adjacent cells not in contact with the pathogen and its toxin retain the ability to express defense reactions. Biotrophic pathogens behave in a more subtle way. They "creep" into the host cell, and establish themselves there but do not kill it. To accomplish this they employ mechanisms which interfere with their recognition by the plant. They bypass or avoid recognition by keeping a "functional distance" from the recognition system, a phenomenon called avoidance. Phytopathogenic bacteria accomplish avoidance by the production of slime which prevents direct contact with the plant. Similarly, *Phytophthora megasperma* f.sp. *glycinea* and *Colletotrichum lindemuthianum* are presumed to escape recognition because the β-glucans in their cell walls, which can act as elicitors for plant defense reactions, are masked by glucoproteins also present in the pathogen's cell wall.

Pathogens may also escape plant defense by suppression of the recognition process. This may occur if a substance produced by the pathogen binds to the receptor and prevents the triggering of defense reactions. Thus the pathogen suppresses its recognition by the plant. A similar process may occur when the expression of race-specific resistance in potato against *Phytophthora infestans* is prevented. The hypersensitive defense reaction (HR) of race-specific resistant potato tuber slices against germinating zoospores of *P. infestans* normally results from triggering the synthesis of the phytoalexin rishitin which kills the leaf tissue. In this case HR is prevented by the production of β-1,3 and β-1,6 glucans when the zoospores germinate. Glucans that act in this way as suppressors have been called hypersensitivity inhibiting factors, or HIFs. The suppressing glucans are produced by *Phytophthora* races with or without mutations to specific virulence (Doke et al.1980). As a consequence, the gene-for-gene incompatibility pattern expected between different pathogen races and different race-specific resistant cultivars (see chapter 9.1.1) is overridden and hidden.

Necrotrophic pathogens do not always instantly kill their hosts after infection. In some cases host cells may stay alive for some time despite the activities of the pathogen. However, if the host plant also carries a resistance gene that enables it to recognize the invading pathogen, it may still express its defense reactions before it is killed. This uncommon resistance response has been observed with *Rhynchosporium secalis*, a necrotrophic pathogen of rye and barley. The build up of secondary defense reactions by a host plant with a race-specific resistance gene forces the pathogen to counter with its pathogenicity factors. In this way a competition or "race" is begun between reaction chains that promote parasitism of the plant and reaction chains to establish new defense barriers. Although such a competition is rare with necrotrophs it is the general rule when biotrophs infect race-specific resistant host plants (see chapter 9.1).

Genes in pathogens that determine phytopathogenicity have been identified experimentally by looking for variants with altered disease symptoms and for non-pathogenic variants still able to grow in axenic culture. Such variants, obtained after mutagen treatment of wild type pathogens, can be crossed with wild type or other mutants with the same or different phenotypes, and the progeny can be scored for segregants of differing phenotypes. The proportions of segregants among phenotypically differing classes allow conclusions about (a) the number of genes involved in expression of a particular phenotype, and (b) the genetic linkage between different mutant loci. Mutants with the same phenotype can be tested in a heterokaryon. If they complement each other, producing a wild type phenotype, this indicates their mutations map in different genes. However, in cases where a heterokaryon test cannot be carried out mixed infection experiments may show whether complementation may occur by diffusion. More biochemically and molecular genetically oriented strategies for the identification of pathogenicity genes will be described in chapter 7.

In mutation experiments with pathogens, several different phenotypes may be expected: mutants defective in pathogenicity that are unable to overcome basic resistance, mutants expressing altered disease symptoms, or mutants with either restricted or extended host range. However, it has been shown that these different phenotypes may be caused by defects in many different genes. All of them may be regarded as pathogenicity genes, even though they may code for quite distinct functions required for parasitism (see also chapt. 7). Still another mutant phenotype can be expected which does not depend on an altered pathogenicity gene. This would be a mutation to specific virulence allowing the mutant pathogen to ignore a so far unrecognized host plant gene for race-specific resistance (see chapt. 9.1.1). The phenotype of specific virulence looks like an extension of the pathogen's host range (see also chapt. 9.1.3.1)

While each pathogen may carry many different pathogenicity genes, only a fraction of them should be involved in determining the pathogen's host range. However, our knowledge of the particular biochemical and physiological functions required by a pathogen to parasitize its host plant is still very fragmentary. This is particularly true for the biotrophs, and for the functions determining the host specificity of a pathogen. Therefore, from a biochemical point of view, one may question whether it is sensible to distinguish between functions determining host range and pathogenicity. For this reason genes determining host range are nowadays ranked among the pathogenicity genes.

Within the last 15 years, molecular genetic methods have made it feasible in a few cases to elucidate the biochemical functions of pathogenicity genes, their genetic structure, DNA sequences, and the nature of their genetic regulation. One of the most important steps toward this goal is to clone the gene and investigate its expression after transformation into a correspondingly defective pathogen, or any other cell suited for observing gene expression. However, the final step in verifying the function of an isolated pathogenicity gene must be the construction, by transformation and gene disruption, of a mutant of the original pathogen defective in the function under consideration. If the defective mutant can be transformed, by homologous recombination, with the isolated wild type pathogenicity gene to the wild pathogenicity phenotype, this will constitute proof that the cloned gene indeed codes for the pathogenicity function in question.

Reviews

Aist J.R., Bushnell W.R. (1991): Invasion of plants by powdery mildew fungi, and cellular mechanisms of resistance. In: Cole G.T., Hoch H.C. (eds.): The Fungal Spore and Disease Initiation in Plants and Animals. Plenum Press, New York, London. 321–345

Bailey J.A. (1983): Biological perspectives of host-pathogen interactions. In: Bailey J.A., Deverall B.J. (eds.): The Dynamics of Host Defence. Academic Press, Australia, Sidney, New York, London, Paris, San Diego, San Francisco, Sao Paulo, Tokyo, Toronto. 1–32

Banuett F. (1992): *Ustilago maydis*, the delightful blight. Trends Genet. 8: 174–180

Boller T (1995): Chemoreception of microbial signals in plant cells. Annu.Rev.Plant Physiol.Plant Mol.Biol. 46: 189–214

Gabriel D.W. (1989): Genetics of plant parasite populations and host-parasite specificity. In: Kosugue T., Nester E.W. (eds.): Plant-Microbe Interactions: Molecular and Genetic Perspectives. Macmillan, New York. 343–379

Goodman R.N., Novacky A.J. (1994): The Hypersensitive Reaction in Plants to Pathogens, A Resistance Phenomenon. APS Press, St.Paul, Minnesota

Hardham A.R. (1992): Cell biology of pathogenesis. Annu.Rev.Plant Physiol.Plant Mol.Biol. 43: 491–526

Heath M.C. (1987): Evolution of plant resistance and susceptibility to fungal invaders. Can. J. Plant Pathol. 9: 389–397

Heath M.C. (1991): Evolution of resistance to fungal parasitism in natural ecosystems. New Phytol. 119: 331–343

Köller W. (1991): The plant cuticle. A barrier to be overcome by fungal plant pathogens. In: Cole G.T., Hoch H.C. (eds.): The Fungal Spore and Disease Initiation in Plants and Animals. Plenum Press, New York, London. 219–246

Kolattukudy P.E., Podila G.K., Sherf B.A., Bajar M.A., Mohan R. (1991): Mutual triggering of gene expression in plant-fungus interactions. In: Hennecke H., Verma D.P.S. (eds.): Advances in Molecular Genetics of Plant-Microbe Interactions. Dordrecht, Kluwer Academic Publishers, Netherlands. 242–249

Kombrink E., Somssich I.E. (1995): Defense response of plants to pathogens. Adv.Bot.Res. **21**: 1–34

Kubo Y., Furusawa I. (1991): Melanin biosynthesis – prerequisite for successful invasion of the plant host by appressoria of *Colletotrichum* and *Pyricularia*. In: Cole G.T., Hoch H.C. (eds.): The Fungal Spore and Disease Initiation in Plants and Animals. Plenum Press, New York and London. 205–218

Mendgen K., Deising H. (1993): Infection structures of fungal plant pathogens – a cytological and physiological evaluation. New Phytol. **124**: 193–213

Schäfer W. (1994): Molecular mechanisms of fungal pathogenicity to plants. Annu.Rev. Phytopathol. **32**: 461–477

Scheffer R.P. (1991): Role of toxins in evolution and ecology of plant pathogenic fungi. Experientia **47**: 804–811

Sprecher E., Urbasch I. (1984): Wechselwirkungen zwischen Pflanzen und pathogenen Pilzen. Naturwiss.Rundsch. **37**: 401–407

Valent B., Chumley F.G. (1991): Molecular genetic analysis of the rice blast fungus, *Magnaporthe grisea*. Annu.Rev.Phytopathol. **29**: 443–467

Yoder O.C., Turgeon B.G. (1994): Molecular determinants of the plant/fungus interaction. In: Kohomoto K., Yoder O.C. (eds.): Host Specific Toxin: Biosynthesis, Receptor and Molecular Biology. Faculty of Agriculture, Tottori University, Sogo Printing and Publishing Co., Ltd., Tottori, Japan. 23–32

Relevant papers

Doke N., Garas N.A., Kuc J. (1980): Effect on host hypersensitivity of suppressors released during germination of *Phytophthora infestans* cytospores. Phytopathology **70**: 35–39

Kimura N., Tsuge T. (1993): Gene cluster involved in melanin biosynthesis of the filamentous fungus *Alternaria alternata*. J.Bacteriol. **175**: 4427–4435

Kubo Y., Nakamura H., Kobayashi K., Okuno T., Furusawa I. (1991): Cloning of a melanine biosynthetic gene essential for appressorial penetration of *Colletotrichum lagenarium*. Molec.Plant-Microbe Interact. **4**: 440–445

Lange L. (1992): Microbes and microbial products in plant protection. Progr.Botany **53**: 252–270

Moore P.D. (1987): Distribution of mycorrhiza throughout the british flora. Nature **327**: 100

Murphy A.M., Pryce-Jones E., Johnstone K., Ashby A.M. (1997): Comparison of cytokinin production *in vitro* by *Pyrenopeziza brassicae* with other plant pathogens. Physiol.Mol. Plant Path. **50**: 53–65

Raggi V. (1987): Water relations in Peach leaves infected by *Taphrina deformans* (Peach leaf curl) – diffuse resistance, total transpiration and water potential. Physiol.Mol.Plant Path. **30**: 109–120

Tosa Y. (1996): Gene-for-gene relationships in *forma specialis-genus* specificity of cereal powdery mildews. In: Mills D., Kunoh H., Keen N.T., Mayama S. (eds.): Molecular Aspects of Pathogenicity and Resistance: Requirement for Signal Transduction. The American Phytopathological Society, St. Paul, Minnesota. 47–55

3.1
Putative Evolution of Fungal Phytopathogenicity

The evolution of fungal pathogenicity probably began with the exploitation of chance contacts between fungi and either the surfaces of plants that offered niches rich in nutrients, or the intercellular spaces of leaves accessible through the stomata. Some fungi may have passively overcome the boundary tissues, epidermis or bark, through accidental wounding. The hypothetical starting point for the evolution of phytopathogenicity may have been the invasion of a plant by a pathogen unable to detect stomata or actively penetrate the epidermis of a leaf or stem.

Plant-pathogen interactions may have progressed through four hypothetical stages. This supposition is based on observations of the different modes of plant-pathogen interaction and their increasing specialization. However, this is not to say that each pathogen existing today should fit exactly into one of these stages of development, or that each highly specialized pathogen must have passed through all four evolutionary stages.

Starting from heterotrophic, non pathogenic fungi, the four different stages may be described as follows:

(1) Pure saprophytes living only on dead organic matter and unable to attack living plants and colonize or parasitize them.

(2) Saprophytes behaving as opportunistic pathogens represent probably the first step towards pathogenicity. Such fungi can colonize living plants, especially those that are debilitated, stressed or senescent. They exhibit only weak virulence. As a rule, invasion occurs through accidental wounds of the leaf epidermis or the bark of a stem. Only the germ tubes of a few more advanced pathogens may find, more or less accidentally, the stomata on a leaf as an entrance to the intercellular space. Alternatively the hyphae form special penetration structures, the so called appressoria, attaching to the cuticle of the epidermis. In this way the pathogen gains, by its own activities, entrance to the plant and its cells. Such opportunistic pathogens kill their host cells and feed on the contents of the dead cells which they digest with exoenzymes. Opportunistic pathogens have little host specificity and plants defend themselves against them by their basic resistance which as stated above may become considerably reduced in weakened or senescent tissue. Examples of opportunistic pathogens include the genera *Cochliobolus, Alternaria, Fusarium* and *Botrytis*.

(3) Primary plant pathogens with only occasional saprophytic feeding under special conditions represent a further stage in the specialization of plant-pathogen interactions. The necrotrophs belong to this group of plant pathogens which, when depending only on saprophytic nutrition on dead and decaying plant material, are not competitive in growth with wholly saprophytic fungi (see stage 1). The long lived reproductive units (for example the chla-

mydospores of the genus *Verticillium* and the sclerotia of *Sclerotinia*) are formed inside the living plant and overwinter in the dead plant material. Specialization within the necrotrophs results in such distinctive features as: formation of special penetration structures, pronounced host specificity and the development of different mechanisms to overcome, bypass, or suppress basic resistance. Many species produce extracellular enzymes like pectinases, cellulases, cutinases, proteases, and toxins which kill their host plants. Pathogen virulence varies according to the activity of the toxin synthesized. While most of the non-host-selective toxins are active against a broad spectrum of plant and animal cells, the action of host-selective toxins is restricted to a single plant species or even to only one mutant carrying specifically sensitive receptors. If the plant lacks such receptors, or can detoxify the toxin, there may be no poisonous effects on the plant. However, this does not mean that the pathogen is necessarily unable to parasitize the plant (for further details see chapters 8.2 and 8.3). – Many pathogenic fungi employ necrotrophic strategies. Examples include the genera *Fusarium, Septoria* and *Pseudocercosporella*. The evolution of toxins and their specificities has been especially well followed within the genera *Cochliobolus* and *Alternaria*.

(4) Biotrophic pathogens represent the most specialized life style among phytopathogenic fungi. Biotrophs derive nutrients from the active metabolism of their host plants. Killing their host cells immediately after infection, like necrotrophs, would be suicidal since they would destroy their food source. Whereas most biotrophs can be cultivated in axenic media, if necessary supplemented with certain vitamins or special nutrients, obligate biotrophs can not be cultivated in artificial media since they seem to have an absolute requirement for nourishment by a living and metabolizing host cell. Examples are nearly all of the rust fungi, *Puccinia* spp. and powdery mildew, *Erysiphe graminis*, both parasites of cereals, downy mildew, *Bremia lactucae*, on lettuce and *Plasmopara viticola* on vine. The entire life cycle of an obligate biotroph can proceed only within living host cells. The obligate biotroph's mode of nutrition thus exploits an ecological niche that allows the pathogen to dispense with some of its own energy consuming synthetic activities. It relies instead on the synthetic capacities of its host.

The plant-pathogen interrelationships within stages 3 and 4 are highly specialized, but based on quite different feeding strategies. Whether necrotrophy and biotrophy developed entirely independently of each other is not known. Indeed transitions from one strategy to another have been found. For example after infection pathogens of the genera *Phytophthora* or *Peronospora* pass first through a biotrophic phase and later, after the break down of host plant tissue, switch to a necrotrophic phase. This kind of nutrition has been called facultative biotrophy or hemibiotrophy.

The nature of the activities of pathogenicity genes developed within the different parasite strategies has not yet been investigated systematically. Therefore current ideas about the evolution of phytopathogenicity in fungi are necessarily speculative. It seems likely that the necessity to overcome the basic resistance of very different plant species would have forced the evolution of specialized pathogenicity genes which would as a result narrow the host range for each of the pathogens. A most instructive example of this among the necrotrophs is the evolution of host-specific toxins (see chapter 8.2). The evolution of biotrophy also points to specialization in exploiting the metabolic activities of host cells. Obligate parasitism is the most extreme example of specialization in this direction.

Reviews

Heath M.C. (1991): Evolution of resistance to fungal parasitism in natural ecosystems. New Phytol. 119: 331 – 343

Scheffer R.P. (1991): Role of toxins in evolution and ecology of plant pathogenic fungi. Experientia 47: 804 – 811

Schmiedeknecht M. (1984): Phylogenie und Evolution des obligat biotrophen Parasitismus im Pflanzenreich. Biol. Rundschau 22: 17 – 29

4 Taxonomy of Phytopathogenic Fungi

Taxonomic classification of non-pathogenic and phytopathogenic fungi down to the species level utilizes mainly morphological and biochemical characteristics. However, within the phytopathogenic species an additional classification is employed based on the plant species parasitized by a pathogen. Individuals of a particular species with the same host range, that parasitize only a single or a few plant species, are classified as belonging to the same sub-group, designated as a *forma specialis* (f.sp.; plural: *formae speciales*, f.spp.) of a pathogen species. Since the host species specificity of a *forma specialis* usually depends on several genetic determinants, the so-called host range genes, this property is a very stable genetic trait. It is taxonomically equivalent to a subspecies among plants.

Each *forma specialis* of a phytopathogenic fungus is named after the genus of the plant it parasitizes. If a *forma specialis* can parasitize more than one plant species, by convention, the genus name of the plant from which the pathogen was first isolated is utilized. Thus the stem rust first isolated from wheat (*Triticum*) is called *Puccinia graminis* f.sp. *tritici*, although this pathogen also parasitizes some other grasses. In cases like these, the identity of the various isolates has to be confirmed by separate tests, for example, by cross infection among the different hosts, or by molecular genetic methods such as comparison of DNA-restriction patterns or Southern hybridization. However, the concept of formae speciales only applies to relatively few pathogens such as *Puccinia graminis, Erysiphe graminis,* and *Verticillium oxysporum.* Many other pathogens with races do not have them, for example *Venturia inaequalis, Cochliobolus* spp., *Cladosporium (Fulvia) fulvum, Phytophthora infestans* etc. implying these pathogens parasitize only one particular plant species.

The host species specificity of a *forma specialis* refers to all individuals of a plant species including all its different cultivars or mutants. At the level of *formae speciales* further classification may be applied to various races that differ in their ability to parasitize cultivars, that carry mutations to race-specific resistance. However, expression of this type of host resistance depends on genetic determinants residing in both the host plant as well as the pathogen (Heath 1991, see also Chap. 9, Sect. 9.1). The genetic determinant contributed by the pathogen, its so called race-specificity, may be lost by a single mutation. Therefore, this property is of no use in taxonomic classification of phytopathogenic fungi. However, physiologic race surveys are a valuable taxonomic tool that enables breeders and farmers to determine what resistant cultivars may be useful in a locality.

Reviews

Gabriel D.W. (1989): Genetics of plant parasite populations and host-parasite specificity. In: Kosugue T., Nester E.W. (eds.): Plant-Microbe Interactions: Molecular and Genetic Perspectives. Macmillan, New York. 343 – 379

5 Plant Defense Strategies Against Phytopathogenic Fungi

How do plants defend themselves against attacking phytopathogenic fungi, and which physiological and biochemical processes are involved? This question is equivalent to the question discussed above: how are pathogens able to parasitize a plant?

Parasitism may be prevented by genetically determined defense mechanisms expressed by the plant. These may represent either constitutive defense barriers, that are always present, or their formation is induced by the attacking pathogen in which case defense reaction expression must be preceded by recognition of the pathogen by the plant.

The processes in plants that prevent their colonization by pathogens can be assigned to two different levels of defense, and on each level different pathogen species or races may be affected. The **first level** of defense, the so-called basic resistance, results from the inability of the pathogen to overcome the defense barriers of the plant. There are many different defense mechanisms expressed in basic resistance that are effective against a very broad range of pathogens. Only a few pathogens equipped with appropriate pathogenicity and host range genes are able to overcome basic resistance.

The **second level** of defense may be established in host plants after basic resistance has been overcome by an infecting homologous pathogen. However, an absolute requirement for establishment of this second level of defense is the presence in the plant of genetic determinant(s) for host resistance. This second level of defense is parasite-specific and directed only towards certain races (see next paragraph). Thus, cultivar-specific resistance enables the plant, in spite of its basic compatibility with the pathogen (see Chap. 7), to defend itself selectively against only a particular homologous pathogen. Other homologous pathogens are still able to parasitize this particular cultivar. Cultivar-specific resistance is nearly exclusively observed against biotrophic pathogens. In summary, cultivar-specific resistance represents a highly selective defense reaction of host plants against one particular pathogen or certain races of it. The most prominent properties of basic- resistance (first level resistance) and cultivar-specific resistance (second level resistance) are summarized in Table 2.

Two different forms of cultivar-specific resistance are observed, race-specific or vertical resistance and race-non-specific or horizontal resistance. Race-specific resistant host plants reject only particular pathogen races (see Chap. 9, and

Table 2. A comparison of basic and host resistance, two different defense mechanisms of plants against attack by biotrophic pathogens

	Basic resistance	Host resistance
	(resistance on the **first** level)	(resistance on the **second** level)
Expressed by	Non-host plants	Host plants
Mode of expression	Constitutive or induced	Constitutive or induced
Specificity	Nearly all pathogens affected, except homologous ones	Only one particular pathogen affected or only certain races of it
Genetic determination	Polygenic	Mono- and oligo- or polygenic
Genetic stability	Stable	Unstable or stable

Sects. 9.1 to 9.1.8), whereas race-non-specific resistant host plants reject all races of one pathogen or *forma specialis*, although not as effectively as in race-specific resistance (see Chap. 9 and Sect. 9.2). However, both types of host resistance are based on quite different defense mechanisms. Furthermore, vertical resistance may be overcome by a single mutation in the pathogen making this type of host plant resistance a rather unstable property. In contrast, horizontal resistance can not be overcome in this way. It is much more stable, at least within the foreseeable agricultural future, because overcoming horizontal resistance would require many mutations in the pathogen. Vanderplank coined the terms "vertical" and "horizontal" resistance. The names were derived from comparing the resistance reactions observed on each of several different cultivars against the same set of pathogen races. Two types of response were found: (1) some cultivars showed very high resistance to some pathogen races of the set, whereas others were fully susceptible. Vanderplank described this as a "vertical" distribution of resistance against different pathogen races. (2) other cultivars exhibited somewhat similar intermediate degrees of resistance against the set of infecting pathogen races. Neither complete resistance nor full susceptibility was observed among the different combinations. The distribution of resistance to different pathogen races was "horizontal". In other words high resistance is confined to particular pairs of cultivar and pathogen race, the defense is race specific, genetically unstable and is called vertical resistance. In contrast, horizontal resistance is not as effective, may vary somewhat, is observed among all pairs of cultivar and pathogen race, is race-non-specific, and genetically very stable. The genetic determination of both types of cultivar-specific resistance is very different. – The terms race specific and race-non-specific resistance are preferable to vertical and horizontal resistance because they are more precise.

Although the release by pathogens of active basic resistance in non-host plants and of defense reactions in race-specific resistant host plants belongs to different levels of plant defense against pathogen attack, the defense mechanisms expressed are very similar in nature.

What do these inducible defense mechanisms look like and what is their physiological and biochemical background? In the past, attention was mostly focused on analyzing race-specific resistance for two experimental reasons: First, their defense mechanisms may be turned on at will following challenge with homologous pathogens or equivalent elicitors and the defense reactions can be quantified. Second, race-specific resistance proved to be relatively easy to analyze, was genetically clearly defined, and near isogenic, race-specific resistant and susceptible host plant lines were available together with more or less homogeneous avirulent and specific virulent (see Chap. 9, Sect. 9.1.1) pathogen populations. These features provided for reliable and easily reproducible experimental results. Since the defense mechanisms triggered in active basic resistance and race-specific resistance employ the same repertoire of gene activities, and since both inducing mechanisms are, as recent results suggest (see Chap. 9, Sect. 9.1.4), rather similar, it seemed justified to utilize experimental results obtained from both types of induced defense for designing models of function.

The defense mechanisms released by recognition of attacking pathogens are the result of signal transduction events that effect activation of genes synthesizing proteins or enzymes engaged in defense reactions. The roles of many of these genes have been elucidated in terms of their physiology and biochemistry. However, a very specific sequence of reactions in the plant's cell membrane occurs (see also Chap. 9, Sect. 9.1.3.3) between pathogen recognition and activation of defense genes that represents part of a signal transduction chain. There are two fundamentally different types of triggered defense reactions in the plant: The first, the so-called *h*ypersensitive defense *r*eaction (HR) of the cell(s) more or less directly affected by the pathogen, occurs very rapidly terminating with plant cell(s) death. The second, which is characteristic of all other inducible defense reactions, develops more slowly in the surrounding and surviving plant cells is sometimes called a normosensitive defense reaction (Klement 1982).

The HR is mainly observed during incompatible interactions between biotrophic pathogens and their race-specific resistant host plants. It is characterized by the rapid death of the penetrated plant cell and of cells in the immediate neighborhood. Hypersensitive cell death occurs a few hours up to one day after pathogen attack, and can be seen at the site where the fungus penetrated as a small, macroscopically recognizable and clearly delimited necrotic fleck. However, in some rare cases in which the HR is exceptionally fast or intense, no necrotic fleck develops since only the penetrated cell, and at most a few neighboring cells, suffer hypersensitive death – too few to be visible macroscopically. The HR has also been observed as a defense mechanism in active basic resistance.

The biochemical and molecular basis of hypersensitive death of plant cells became clear only recently. The steps leading to it may be described as follows: The recognition between host plant and its pathogen takes place at the plant cell membrane. A chemical signal is generated which is transmitted by a signal transduction reaction chain within the cell membrane that employs enzymes and second messengers (see Chap.9, Sect. 9.1.3.3), finally leading to expression of hypersensitive death. Most importantly, triggering of hypersensitive cell death requires active cell metabolism and is determined and regulated by a genetic program residing in the attacked plant cell. Therefore one speaks of a *pro-grammed cell death* (PCD; see also Chap. 9, Sect. 9.1.4).

As a consequence of the hypersensitive death of the plant cell, the penetrating pathogen loses its food source and becomes surrounded by a barrier of hypersensitively killed plant cells, thereby forcing the pathogen to death by starvation. In this way the plant sacrifices parts of its own tissue together with the intruding parasite. The signals generated by the recognition event and transduced by signals via the plant cell membranes may also trigger transcription of genes coding for the second type of defense reaction in neighboring surviving and metabolizing plant tissue. These normosensitive reactions are expressed much later than the HR and consist of the erection of newly expressed defense barriers; which will now be discussed more fully.

Normosensitive defense reactions of living and metabolizing cells include reinforcement of cell walls by attachment of callose, formation of papillae preventing penetration of the pathogen, thickening of the cuticle, and other kinds of cell wall reinforcement. This type of defense reaction, triggered by recognition between host and pathogen, may occur either in close proximity to a hypersensitively killed cell or in tissue with no accompanying hypersensitive death. Many of the triggered normosensitive defense reactions are based on the activation of genes coding for enzymes of phenyl-propanoid metabolism, e.g., phenylalanine ammonialyase (PAL) and 4–coumarate:CoA ligase (4CL), and of genes coding for enzymes synthesizing subsequent products such as the phytoalexins (Mieth et al. 1986; Rohwer et al. 1987; Fritzemeier et al. 1987). Finally, the products engaged in pathogen defense appear to spread systemically over the whole plant, or at least parts of it, as has been observed for chitinases and glucanases that are able to digest the cell walls of attacking phytopathogenic fungi or bacteria (Broglie et al. 1991; Jach et al. 1992; Fluhr et al.1991). However, it is not the enzymes that spread systemically but rather signal substances able to trigger their synthesis. This phenomenon of spreading resistance which protects plants against most of their different pathogens has been called *systemic acquired resistance* (SAR) or induced resistance (see Chap. 11).

Proteins and enzymes first appearing during pathogen defense, i.e., after recognition of an attacking pathogen by the resistant plant, are synthesized *de novo*, since the appearance of newly synthesized, corresponding mRNA species was also demonstrated. Hence, these proteins and enzymes do not originate from activation of already present precursors. This has been shown in experiments employing intact plants as well as single-cell cultures of plants or protoplast preparations. The newly synthesized proteins were called defense related proteins. The corresponding coding sequences were in many cases identified on the basis of their function in pathogen defense, although – as it turned out later – many of them are also temporarily expressed during normal plant development, i.e., under normal conditions of developmental regulation and tissue-specific expression.

A sub-group within these defense-related proteins consists of the so-called pathogenesis related or PR-proteins. These were first recognized in TMV (tobacco mosaic virus)-infected, TMV-resistant tobacco cultivars exhibiting a hypersensitive defense reaction. They were later found in intercellular fluids of other race-specific resistant plants infected with different phytopathogens, or after treatment of plants with biotic or abiotic elicitors. Many of the PR-proteins obtained from intercellular fluids are located in cell walls and cellular vacuoles. Some PR-proteins are also found after treatment of plants with chemicals such as polyacrylic acid, metal salts, salicylic acid, or phytohormones such as auxin, cytokinin or ethylene, and finally also under stress conditions e.g., osmotic or salt stress. Five different "families" of PR-proteins (PR-1 to PR-5) have been identified by polyacrylamide gel electrophoresis (PAGE).The proteins in each family have similar properties such as molecular mass, serologic relationship, nucleotide and amino acid sequences, enzymatic activities, acidic and basic isoforms. Each of these families are coded for by several genes; however, originally their functions were unknown. Only with increasing numbers of identified PR-proteins could defined enzymatic functions be assigned to them, for example, β-1,3 – glucanases, chitinases or α-amylases, and proteinase inhibitors. Since similar functions were already previously assigned to some of the defense related proteins, the distinction between PR-proteins and defense – related proteins has become rather vague, and may indeed become pointless in the future as PR-proteins are classified according to their functions. However, PR-proteins remain of special interest as an experimentally easily accessible and well-defined group of proteins among the defense related proteins. As already mentioned, many of the proteins participating in pathogen defense are also expressed during the normal developmental regulation of the plant. This suggests that during evolution plants have adapted existing genes for performing newly required functions such as pathogen defense or adapting to other stress situations.

The following short survey of induced defense-related proteins refers only to what is known so far of their functions in pathogen defense, but not to the methods employed in isolating them, i.e., without linking them to the PR-proteins. Two groups may be established:

The first group comprises structural proteins and enzymes that strengthen, alter, and repair the plant's apoplast. The apoplast is the continuum represented by the plant's cell walls and bordering on the interior of cells; it is comprised of many different classes of substances. The apoplast plays a central role in the defense strategies of plants, and several different protein species are built into the cell walls: These include structural proteins, such as hydroxyproline-rich gluco-proteins (HRGP, extensins) (Corbin et al. 1987) and glycine-rich proteins (GRP), and enzymes such as peroxidases, engaged in the alteration or *de novo* synthesis of the cell wall polymers lignin and suberin, as well as those synthesizing oxidized phenols effecting enzyme inactivation. Reinforcement of plant cell walls is ac-complished by callose synthase, which catalyzes the synthesis of β-1,3 glucan. This compound seals off cell walls at sites of attempted pathogen penetration. Callose formation is a very rapidly expressed defense reaction which is easily detected morphologically as newly formed papillae. HRGPs function in pathogen defense by becoming cross-linked by the action of reactive oxygen species pro-duced during the oxidative burst induced by pathogen attack. This makes the cell wall more refractory to enzymatic digestion. Cross linking is a very fast reaction and occurs even when transcription or translation is blocked in the plant.

The second group of proteins comprises enzymes not bound to the apoplast and which are engaged in synthesizing antimicrobial substances such as lectins and thionins. This group also includes inhibitors such as amylases, proteinases, and thaumatins, as well as endohydrolases, e.g., β-1,3 glucanases and chitinases, and enzymes taking part in synthesizing tannins, o-chinons and the low mole-cular weight phytoalexins (the latter will be discussed below in more detail). β-1,3 glucanases digest cell walls of penetrating pathogens thereby liberating glucan fragments which then act as exogenous elicitors for triggering plant defense re-actions. Furthermore, plant-borne chitinases are able to disassemble the chitin polymer of the cell walls of intruding pathogens. The synthesis of chitinases by the plant requires induction by the glucan fragments liberated by the plant-borne β-1,3 glucanases from the glucans embedded in the chitin polymer of the patho-gen. Hence the induced synthesis of glucanases and chitinases in plants has to follow a fixed sequence. Peroxidases should be mentioned here because, besides their function in lignin formation, they capture active oxygen compounds, in-cluding the superoxide radical O_2^-, the peroxide radical HO_2 and its product H_2O_2, all of which arise from the interaction of the plant with pathogens and which are very harmful to plant cells.

The phytoalexins, synthesized by enzymes belonging to the second group of defense related proteins, are low molecular weight antimicrobial substances. They are derived from different classes of compounds such as phenols, tannins, flavonoids, terpenoids, or cyanogenic glucosides. Their appearance may be de-tected within a few hours, or at the most 1 or 2 days, following pathogen recog-nition and the hypersensitive reaction. Under such conditions most dicots form phytoalexins, whereas among monocots so far no synthesis of phytoalexins has been observed except the formation of the phytoalexin avenalumin, synthesized by oat plants after infection with rust fungi.

Most phytoalexins are active within broad pH and temperature ranges and their antibiotic activities are directed unspecifically at pathogens as well as plant cells, provided they are present in sufficiently high concentrations. Phytoalexins are products of the host plant's secondary metabolism. Abiotic stresses such as mechanical wounding, UV irradiation, or the action of certain chemicals, may also trigger phytoalexin synthesis, which in some cases can reach antibiotic concentrations. However, the fungicidal activity of phytoalexins *in vivo* has been demonstrated unequivocally only for the pisatin of pea plants during expression of active basic resistance. After infection of pea plants with *Nectria haematococca*, a rapid onset of pisatin synthesis is induced reaching concentrations sufficient to kill the invading pathogen. However, to counteract this killing effect the pathogen may synthesize a pathogenicity factor which detoxifies pisatin. The pathogenicity factor is a P-450 monooxygenase which demethylates pisatin rendering it harmless. A mutant of *N. haematococca* defective in the gene coding for the monooxygenase was unable to parasitize pea plants. The example demonstrates that a pathogenicity factor, the monooxygenase, overcomes a defense mechanism belonging to active basic resistance, to establish basic compatibility between the pathogen and its host plant.

Meanwhile it became obvious that the synthesis and action of phytoalexins does not bring about the necrosis accompanying a hypersensitive reaction. Rather, the HR sets in motion in the penetrated cell and in neighboring tissue a genetic program which regulates expression of defense genes. Among them are genes engaged in programmed cell death, PCD, as well as phytoalexin synthesis (see also Chap. 9, Sects. 9.1.3.1, 9.1.3.3 and Sect. 9.1.4).

Phytoalexin synthesis may, as discussed above, also be induced during the expression of race-specific resistance, i.e., under conditions of basic compatibility, during which the homologous interaction becomes incompatible because the host plant is race-specifically resistant. However, even during compatible interactions in the absence of race-specific resistance the synthesis of phytoalexins may be induced. Typically, under these conditions, phytoalexin synthesis is much lower, reaches its peak later, and does not hinder colonization. This points to a general phenomenon observed in plant-pathogen interactions, namely, competition inside the plant between two opposing reaction chains. One originates from the pathogen and is aimed at establishing stable colonization of the host plant, whereas the other originates from the plant and is directed at preventing parasitism. When defense against the pathogen wins, as in incompatible interactions of pathogens with race-specific resistant plants, its success is due to its occurring faster and more effectively, not to turning off the expression of the parasite's pathogenicity genes. If the pathogen "escapes" recognition by its host plant and thereby parasitizes it, the same principle is valid: the pathogen is successful because its pathogenicity genes acted faster than the plant's defense genes (see also Chap 9, Sect. 9.1).

Induction of synthesis of several phytoalexins with more or less similar chemical structures is observed when the phytoalexins are coded by gene families. Phytoalexins may vary from one species to another and related plants often synthesize similar phytoalexins. Also, differences in quantity among similar phytoalexins are found within the same plant species, and they may vary with the type of tissue or with the inducing agent, for example, a phytopathogenic fungus or an insect.

In the past, the significance of phytoalexins in plant defense reactions has been somewhat overestimated. In the meantime, it has become apparent that phyto-alexin formation represents only one component within an array of defense mechanisms. However, in many cases of plant-pathogen interactions, synthesis of phytoalexins is indicative of the release of defense reactions. Accordingly, the temporal progress of defense reactions may be followed by measuring the rates of phytoalexin synthesis.

In contrast to phytoalexins, determined by gene families coding for structu-rally somewhat different products, the synthesis of other defense products, such as callose, appear as uniform products and the same holds for enzymes active in defense like chitinases or β-1,3 glucanases. Any structural variations observed among such products originate from expression of different individual compo-nents of basic resistance. In such cases either a single elicitor induces several components of active basic resistance, or the pathogen presents several elicitors to the plant, and each elicitor is able to induce a particular component of basic resistance.

All of the defense reactions listed above are characteristic of the plant species and infecting pathogen, and each reaction may vary in intensity. The reaction also may be modulated qualitatively and quantitatively according to the devel-opmental state of the plant (Fluhr et al. 1991) and environmental conditions. There is also a finely-tuned coordination between different defense reactions, as the investigation of hypersensitive reactions in race-specific resistance reac-tions has disclosed. Thus there is coordination in time, when enzyme activities are induced in a distinct sequence, and in space, when some activities are trig-gered close to the site of infection, and others occur more distantly. Cell and tissue specificity, is also observed, with certain activities only being expressed by particular cell types.

Some of the biochemical activities induced by pathogen attack also produce morphological and cytological changes in the plant which are easily observable. Examples include the aggregates of cytoplasm which appear in some cells as a prelude to a defense reaction, or the formation of halos and callose papillae at sites of pathogen penetration. Currently, only some of these morphological struc-tures can be linked to one or more biochemical activities, such as incorporation of lignin or suberin into plant cell walls. These gaps in knowledge will probably be closed in the near future as more refined experimental techniques become available.

In summary, it can be stated that the hypersensitive defense reaction, like other defense reactions, is the result of many cellular and molecular events. These are triggered by one or several elicitors which, via cascades of signal transduction events involving as yet unknown signal substances, effect the observed defense reactions. Their expression seems, in most cases, to be of a local nature, but may also spread systemically throughout the plant by means of signal transduction cascades and amplification events (see also Chap. 11, Induced Resistance).

Reviews

Aist J.R. (1983): Structural responses as resistance mechanisms. In: Bailey J.A., Deverall B.J. (eds.): The Dynamics of Host Defence. Academic Press Australia, Sidney, New York, London, Paris, San Diego, San Francisco, Sao Paulo, Tokyo, Toronto. 33–70

Aist J.R., Bushnell W.R. (1991): Invasion of plants by powdery mildew fungi, and cellular mechanisms of resistance. In: Cole G.T., Hoch H.C. (eds.): The Fungal Spore and Disease Initiation in Plants and Animals. Plenum Press, New York, London. 321–345

Bailey J.A. (1983): Biological perspectives of host-pathogen interactions. In: Bailey J.A., Deverall B.J. (eds.): The Dynamics of Host Defence. Academic Press Australia, Sidney, New York, London, Paris, San Diego, San Francisco, Sao Paulo, Tokyo, Toronto. 1–32

Barz W. (1997): Phytoalexins. In: Hartleb H., Heitefuss R., Hoppe H.-H. (eds.): Resistance of Crop Plants against Fungi. Gustav Fischer, Jena, Stuttgart, Lübeck, Ulm. 183–201

Boller T. (1995): Chemoreception of microbial signals in plant cells. Annu.Rev.Plant Physiol.Plant Mol.Biol. **46**: 189–214

Bowles D.J. (1990): Defense-related proteins in higher plants. Annu.Rev.Biochem. **59**: 873–907

Crute I.R., de Wit P.J.G.M., Wade M. (1985): Mechanisms by which genetically controlled resistance and virulence influence host colonization by fungal and bacterial parasites. In: Fraser R.S.S. (ed.): Mechanisms of Resistance to Plant Diseases. Martinus Nijhoff/ Dr.W.Junk, Dordrecht, Boston, Lancaster. 197–309

Cutt J.R., Klessig D.F. (1992): Pathogenesis related proteins. In: Meins F., Boller T. (eds.): Plant Gene Research,. Springer Verlag, New York. 209–243

Dixon R.A. (1986): The phytoalexin response: Elicitation, signalling and control of host gene expression. Biol.Rev. **61**: 239–291

Dixon R.A., Harrison M.J. (1990): Activation, structure, and organization of genes involved in microbial defense in plants. Adv.Genetics **28**: 165–234

Ebel J., Grisebach H. (1988): Defense strategies of soybean against the fungus *Phytophthora megasperma* f.sp. *glycinea*: a molecular analysis. Trends Biochem.Sci. **13**: 23–27

Freytag S., Hahlbrock K. (1992): Abwehrreaktionen von Pflanzen gegen Pilzbefall. Biologie in unserer Zeit **22**: 135–142

Goodman R.N., Novacky A.J. (1994): The Hypersensitive Reaction in Plants to Pathogens, A Resistance Phenomenon. APS Press, St.Paul, Minnesota

Hammond-Kosack K.E., Jones J.D.G. (1996): Resistance Gene-Dependent Plant Defense Responses. Plant Cell **8**: 1773–1791

Heath M.C. (1991): Evolution of resistance to fungal parasitism in natural ecosystems. New Phytol. **119**: 331–343

Keen N.T. (1992): The molecular biology of disease resistance. Plant Mol.Biol. **19**: 109–122

Keen N.T. (1993): An overview of active disease defense in plants. In: Fritig B., Legrand M. (eds.): Mechanisms of Plant Defense Responses. Kluwer Academic Publishers, The Netherlands. 3–11

Klement Z. (1982): Hypersensitivity. In: Mount M.S., Lacy G.H. (eds.): Phytopathogenic Prokaryots. Academic Press, Inc., New York, London, Paris, San Diego, Sao Paulo, Sidney, Tokyo, Toronto. 149–177

Kolattukudy P.E., Podila G.K., Sherf B.A., Bajar M.A., Mohan R. (1991): Mutual triggering of gene expression in plant-fungus interactions. In: Hennecke H., Verma D.P.S. (eds.): Advances in Molecular Genetics of Plant-Microbe Interactions. Dordrecht, Kluwer Academic Publishers, Netherlands. 242–249

Kombrink E., Somssich I.E. (1995): Defense response of plants to pathogens. Adv.Bot.Res. **21**: 1–34

Linthorst H.J.M. (1991): Pathogenesis-related proteins of plants. Crit. Rev. Plant Sci. **10**: 123–150

Moerschbacher B.M., Reisener H.-J. (1997): The hypersensitive resistance reaction. In: Hartleb H., Heitefuss R., Hoppe H.-H. (eds.): Resistance of Crop Plants against Fungi. Gustav Fischer, Jena, Stuttgart, Lübeck, Ulm. 126–158

Robinson R.A. (1973): Horizontal resistance. Review of Plant Pathology **52**: 483–501

Rushton P.J., Somssich I.E. (1998): Transcriptional control of plant genes responsive to pathogens. Curr. Opinion Plant Biol. **1**: 311–315

Ryan C.A. (1990): Protease inhibitors in plants: Genes for improving defenses against insects and pathogens. Annu.Rev.Phytopathol. **28**: 425–449

Smart M.G. (1991): The plant cell wall as a barrier to fungal invasion. In: Cole G.T., Hoch H.C. (eds.): The Fungal Spore and Disease Initiation in Plants and Animals. Plenum Press, New York and London. 47–66

Stinzi A., Heitz T., Prasad V., Wiedemann-Merdinoglu S., Kauffmann S., Geoffroy P., Legrand M., Fritig B. (1993): Plant 'pathogenesis-related' proteins and their role in defense against pathogens. Biochimie **75**: 687–706

Stoessel A. (1983): Secondary plant metabolites in preinfectional and postinfectional resistance. In: Bailey J.A., Deverall B.J. (eds.): The Dynamics of Host Defence. Academic Press Australia, Sidney, New York, London, Paris, San Diego, San Francisco, Sao Paulo, Tokyo, Toronto. 71–122

Scholtens-Thoma I.M.J., Joosten M.H.A.J., de Wit P.J.G.M. (1991): Appearance of pathogen-related proteins in plant hosts – Relationships between compatible and incompatible interactions. In: Cole G.T., Hoch H.C. (eds.): The Fungal Spore and Disease Initiation in Plants and Animals. Plenum Press, New York, London. 247–265

Tenhaken R., Levine A., Brisson L.F., Dixon R.A., Lamb C. (1995): Function of the oxidative burst in hypersensitive disease resistance. Proc.Natl.Acad.Sci.USA **92**: 4158–4163

Tzeng D.D., DeVay J.E. (1993): Role of oxygen radicals in plant disease development. Advances in Plant Pathology **10**: 1–34

Vanderplank J.E. (1968): Disease Resistance in Plants. Academic Press, New York, London

Wubben J.P., Boller T., Honée G., De Wit P.J.G.M. (1997): Phtoalexins. In: Hartleb H., Heitefuss R., Hoppe H.-H. (eds.): Resistance of Crop Plants against Fungi. Gustav Fischer, Jena, Stuttgart, Lübeck, Ulm. 202–237

Relevant papers

Broglie K., Chet I., Holliday M., Cressman R., Biddle P., Knowlton S., Mauvais C.J., Broglie R. (1991): Transgenic plants with enhanced resistance to the fungal pathogen *Rhizoctonia solani*. Science **254**: 1194–1197

Fluhr R., Sessa G., Sharon A., Ori N., Lotan T. (1991): Pathogenesis-related proteins exhibit both pathogen-induced and developmental regulation. In: Hennecke H., Verma D.P.S. (eds.): Advances in Molecular Genetics of Plant-Microbe Interactions. Dordrecht, Kluwer Academic Publishers, Netherlands. 387–394

Fritzemeier K.-H., Cretin C., Kombrink E., Rohwer F., Taylor J., Scheel D., Hahlbrock K. (1987): Transient induction of phenylalanine ammonia-lyase and 4-coumarate:CoA ligase mRNA in potato leaves infected with virulent or avirulent races of *Phytophthora infestans*. Plant Physiol. **85**: 34–41

Jach G., Logemann S., Wolf G., Oppenheim A., Chet I.S. J, Logemann J. (1992): Expression of bacterial chitinase leads to improved resistance of transgenic tobacco plants against fungal infection. Biopractice **1**: 33–40

Mieth H., Speth V., Ebel J. (1986): Photoalexin production by isolated soybean protoplasts. Z.Naturforsch. **41c**: 193–201

Rohwer F., Fritzemeier K.-H., Scheel D., Hahlbrock K. (1987): Biochemical reactions of different tissues of potato (*Solanum tuberosum*) to zoospores or elicitors from *Phytophthora infestans*. Accumulation of sesquiterpenoid phytoalexins. Planta **170**: 556–561

6 Basic Resistance: The Absence of Parasitism of Non-Host Plants by Phytopathogenic Fungi

This chapter deals more thoroughly with the features of basic resistance and stresses the contrast with basic compatibility (Chap. 7) which leads to parasitism. Most plants remain free of infection by microbial pathogens because they are non-hosts for them. This may depend on the constitutive expression of defense barriers, such as a particularly thick cuticle protecting the epidermis of leaves, synthesis of poisonous substances, e.g., alkaloids or saponins, and/or a repertoire of different inducible defense mechanisms. On the other hand, the phenomenon of "defense" against a pathogen may be simulated by the parasite's inability to colonize the plant, because it lacks suitable pathogenicity genes. General speaking, in this case matching between the pathogen and its host plant for colonization is lacking. All factors that preclude the plant's colonization by pathogens are included in the term "basic resistance" (non-host resistance, basic incompatibility). Less commonly used but doubtless more appropriate is the term "basic incompatibility," because it comprises both aspects of preventing colonization of the plant: the lack of appropriate pathogenicity genes by the pathogen, and defense by the basic resistance mechanisms of the plant. At times the term immunity has also been used.

The genetic determination of basic resistance is very complex because many components of both pathogen and plant come into play. Correspondingly, basic resistance depends on many different genes. Furthermore, it is directed at dissimilar species of phytopathogenic fungi, bacteria and viruses, at phytophagous arthropods and nematodes, and is genetically a very stable trait (see also Table 2). The reason for the absence of selectivity in defense against different pathogen species is based on the diversity of defense mechanisms of basic resistance in plants. This diversity ensures that the different methods of attack and the various pathogenicity functions employed by different pathogens are counteracted by a number of distinct plant defense mechanisms, some of which are already triggered by mechanical wounding by the attacking pathogen. Wounding leads to the liberation of plant hydrolytic enzymes that digest plant cell wall polymers, forming, as decomposition products, endogenous elicitors which trigger the expression of defense reactions. Purely abiotic effects, such as mechanical wounding, chemicals, or irradiation with ultraviolet light, may also result in the expression of particular defense reactions. However, the lack of pathogen species specificity observed in basic resistance is in stark contrast to the high specificity involved in the recognition function of the non-host plant that triggers the expression of particular defense reactions. These are based on interaction between a particular signal molecule, the elicitor, and a corresponding plant receptor (see Chap. 9, Sect. 9.1.4).

In contrast to the nonselective action of basic resistance, the colonization of a plant by a particular pathogen, i.e. the presence of basic compatibility between both partners, is a highly selective and species-specific event. It is based on the evolution of many different pathogenicity genes in one particular pathogen species, or *forma specialis*, which thus acquires the ability to parasitize one particular host plant.

Generally, the mechanisms of basic resistance may be passive or active in nature. Passive basic resistance is based on mechanisms of constitutive expression which act as the first defense barrier against pathogen attack (see also Table 2). The terms passive and constitutive mean that these mechanisms are permanently present or expressed and include, for example, the morphology characteristic of an especially resistant cuticle or the constitutive synthesis of a toxic substance. Passive basic resistance is the first component of basic resistance and is highly effective against pathogen attack. It has been proposed that constitutively synthesized defense compounds be called "phytoanticipins," as opposed to the phytoalexins, which are induced by pathogen attack. Phytoanticipins are synthesized constitutively either as biologically active compounds, e.g., saponins, or as inactive compounds, e.g., cyanogenic glucosides, which are activated by plant-borne enzymes liberated following destruction of the plant's integrity by an attacking necrotrophic pathogen, in this case by liberation of hydrolases generating hydrogen cyanide from the cyanogenic glycosides.

Having succeeded in overcoming the first defense barrier of basic resistance, the pathogen may now be confronted by a second barrier of active basic resistance, triggered by the cell-to-cell interaction between pathogen and plant. This interaction is always linked to a recognition reaction made feasible by wounding caused either by penetration of a plant cell by the pathogen or by some substance(s) produced by the pathogen. As mentioned already, one essential element of recognition (Chap. 2) is a signal product or elicitor, such as a secretion or surface molecule of the pathogen. Since these not of plant origin, they are called exogenous elicitors. However, among the enormous numbers of different substances produced by fungi, only certain molecules act as elicitors and these may vary with the pathogen as well as with the plant recognizing them as elicitors. We do not yet know the exact prerequisites for elicitor function. If a natural elicitor does not trigger hypersensitive cell death but instead induces all the other defense reactions, it is called a non-HR elicitor.

Among the non-HR elicitors the oligosaccharides play a dominating role. For example, glucan oligosaccharides isolated from cell walls of *Phytophthora megasperma* by enzymatic or chemical hydrolysis, induce the synthesis of phytoalexins in many different plant species (Ham et al. 1991). Oligogalacturonides, also acting as elicitors, are liberated from the pectin skeleton of soybean cell walls by enzymes of the phytopathogenic bacterium *Erwinia caroto*vora (Davis et al. 1984). Protein containing substances as well as molecules derived from lipids also rank among the non-HR elicitors. If all pathogens belonging to the same taxon produce the same elicitor molecule recognized by one or more plant species, one speaks of a general elicitor (see, in contrast, the race-specific elicitor, Chap. 9, Sect. 9.1.3.1).

Consistent with the large number of different phytopathogenic fungi, and many different pathogenicity genes, a wide range of parasite "techniques," may be expected for overcoming or bypassing basic resistance defense mechanisms. Some were discussed in Chap. 3; but three typical examples will be mentioned here. Necrotrophs kill host cells by the action of their extracellular enzymes or by toxins, and prevent them from producing new defense barriers. In other cases pathogen enzymes digest the barriers of passive or induced basic resistance, or they detoxify toxin barriers, as discussed above for the phytoalexin pisatin. Finally, biotrophs may negate active basic resistance by avoiding host plant recognition.

In particular cases, exogenous elicitors of active basic resistance may also be very specific in nature and origin. For example three species of *Phytophthora* each synthesize their own elicitin: *Phytophthora cryptogea* cryptogein, *Phytophthora capsici* capsicein and *Phytophthora parasitica* parasiticein (Ricci et al. 1989, 1992, Blein et al. 1991, Kamoun et al. 1993b). These elicitins are proteins with similar amino acid compositions, are about 10 kDa in size, and are secreted into the medium during cultivation *in vitro*. Elicitins cause necrosis or defense reactions when infiltrated into the leaves of the non-host plant tobacco or when applied to detached shoots. Elicitin-treated tobacco plants also express induced systemic resistance (SAR, see Chap. 11) against infection by other pathogens, among them *Phytophthora parasitica* f.sp. *nicotianae*, the causative agent of black shank disease. This particular *forma specialis* of *P. parasitica* does not produce parasiticein. It is believed that this is why it is able to colonize tobacco plants: Lacking parasiticein, it can no longer trigger active basic resistance, and so tobacco has become its host plant. However, all other members of the species *Phytophthora parasitica* exhibit a very broad host range, but still show preference for certain plant species, such as *Solanum melongena*, *Dianthus caryophyllus*, *Citrus limon*, *Citrus jambhiri* and *Ananas comosus*. In these plants parasiticein releases either no defense reactions or only weak ones. In other words, parasiticein acts on tobacco plants like an avirulence factor, releasing a hypersensitive defense reaction (HR). The fact that *P. parasitica* f.sp. *nicotianae* does not produce parasiticein confers basic compatibility with tobacco. It is supposed that elicitins act primarily as virulence factors, albeit with different efficiencies in different plant species. However, in exceptional cases like tobacco a particular elicitin, parasiticein, may be recognized as an avirulence factor, since this host species produces a corresponding receptor protein.

The genes coding for the synthesis of parasiticein and an elicitin of *P. infestans* have been isolated (Kamoun et al. 1993a,1997). Both produce a protein of 118 amino acids with an amino-terminal signal sequence of 20 amino acids processed for secretion. Genes synthesizing elicitins belong to gene families common to all *Phytophthora* species, and in *P. parasitica* there are at least three member genes. Isolates of *P. parasitica* f.sp. *nicotianae* unable to produce parasiticein neverthe-less contain elicitin coding sequences, but no mRNA is transcribed from them.

To explain the phenomenon of active basic resistance a quite different hypothetical explanation has been proposed, the so-called suppressor-model of basic resistance (Bushnell and Rowell 1981). It assumes that active basic resistance is triggered unspecifically by general acting elicitors (see above) produced by all pathogens and likewise recognized by receptors present in all plants. However, particular pathogens may become compatible with certain plant species because, by virtue of mutation, the pathogen produces a species-specific suppressor that prevents its general elicitor from acting on the plant receptor, thus negating active basic resistance and effecting basic compatibility allowing for colonization of the plant. The suppressor of active basic resistance could, for example, compete with the general elicitor binding to the plant receptor (Yoshikawa and Sugimoto 1993), or block elicitor-receptor interaction in some other way, disturbing subsequent signal transduction, or hindering formation or action of the effector. In short, basic resistance would be prevented by a specific suppressor produced by the pathogen thus allowing basic compatibility. Under such circumstances, a specific pathogen-plant interaction, namely, that of the specific suppressor with the plant receptor, would not trigger pathogen defense, but would instead establish basic compatibility, i.e., pathogen susceptibility. Bailey described this as elicitor/specific suppression, and contrasted it to the release of active basic resistance by a specifically acting elicitor, as discussed at the beginning of this chapter. Similar mechanisms of specific suppression that release basic resistance include the detoxification of pisatin by a demethylase of *Nectria haematococca* (see Chap. 5), or the formation of fast-acting toxins that prevent expression of HR.

Phytophthora infestans provides an example of suppression of elicitation of a defense reaction caused by direct receptor-suppressor binding or by preventing liberation of an effector molecule: Two general elicitors, eicosapentanoic acid and arachidonic acid, were isolated from cell walls of *P. infestans* which triggered the synthesis of sesquipertenoid phytoalexins in potato slices (Bostock et al. 1982). This elicitation of a defense reaction can be specifically suppressed by water-soluble β-1,3 glucan and β-1,6 glucans, produced by the zoospores of the attacking pathogen (Doke et al. 1980). A similar situation has been observed with the pathogen of peas, *Mycosphaerella pinodes*. Its germinating spores secrete, in addition to a high molecular weight general elicitor of more than 10 kDa, two smaller glycopeptides which suppress elicitation of basic resistance, thereby suppressing accumulation of the phytoalexin pisatin. The the two glycopeptides, supprescins A and B (Oku et al. 1994), reduce ATPase activity in the plant cell membrane. This probably blocks signal transduction, i.e., the effector for releasing pisatin synthesis. Thus, *M. pinodes* establishes basic compatibility by suppressing the elicitation of active basic resistance by the general elicitor produced by the same germinating spores.

These examples of the suppression of active basic resistance are evidence of the coevolution between plants and pathogens. Basic resistance is expressed, or elicited, because a pathogen-borne general elicitor interacts with a corresponding receptor in the plant. In a first mutational step, a specific suppressor A', synthesized by the pathogen and active only in plant species A, may come into being, that prevents interaction of the general elicitor with the plant's receptor thereby abolishing expression of basic resistance resulting in basic compatibility with plant species A. As a second step, in the course of evolution of plant species A to species B, a mutation could occur in the plant's receptor so that its interaction with the general elicitor can no longer be prevented by suppressor A', thereby reestablishing active basic resistance. In a third step the pathogen's suppressor A' may acquire by mutation the changed specificity B' enabling it to prevent the plant receptor from interacting with the pathogen's general elicitor, allowing again for expression of basic compatibility and so forth. In this way alternating mutational changes in the plant's receptor for the general elicitor and in the pathogen's species for the specific suppressor might drive this coevolution. However, there is no proof that active basic resistance has evolved in nature according to this model. The examples of suppression of interaction of the *P. infestans* general elicitors eicosapentanoic acid and arachidonic acid by β-1,3- and β-1,6 glucans, and of the high molecular weight general elicitor of *M. pinodes* by a low molecular weight glycopeptide, are exceptional cases, that only demonstrate the feasibility of suppressor action in preventing elicitation of basic resistance.

The nature of basic resistance varies among different plant species. It may be constitutive or induced; it may impede penetration, or result from synthesis of substances toxic for the pathogen; and may result from different elicitors. Each plant or tissue may have its own particular mixture of defense mechanisms. Therefore, each pathogen is confronted with a unique mixture of components of basic resistance. If the pathogen manages to overcome many plant defense mechanisms by employing its battery of pathogenicity genes, then it exhibits a broad host range. It is to be expected that a single pathogen will trigger in its host plant species only a limited number of the possible components of active basic resistance. Furthermore, various plants carrying batteries of components for active basic resistance should harbor the same, or at least very similar, components of active basic resistance for defense against a particular pathogen. These phenomena possibly explain why a single fungus is rarely restricted to a single host species. However, at the same time they also lead one to expect that there should be many more examples of broad host range pathogens that have avoided the actice basic resistance that all those hosts have in common. There may indeed be many such pathogens that go undetected because they are

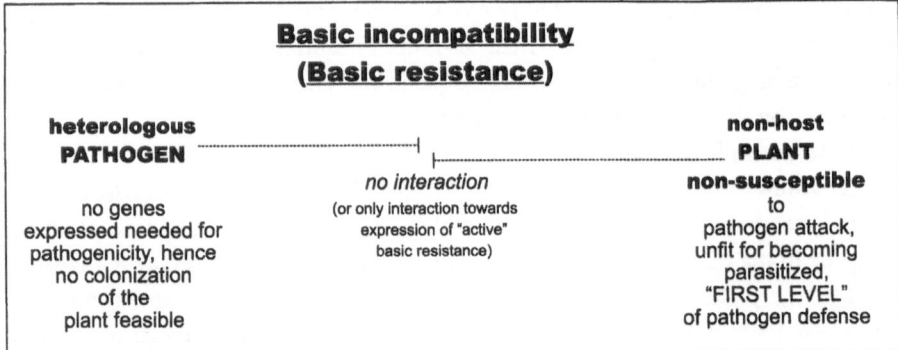

Figure 1. Basic incompatibility between plants and attacking pathogens implies no inter-action occurring between pathogen and plant, or only an interaction resulting in the expression of the defense mechanisms of „active" basic resistance

prevented from colonizing most host plants by effective race specific resistance which we have yet to recognize (see also Chap. 4, Sect. 9.1.3.1). – The main facts defining basic incompatibility and basic resistance are summarized in *Fig. 1.*

Reviews

Atkinson M.M. (1993): Molecular mechanisms of pathogen recognition by plants. Advances in Plant Pathology **10**: 35–64

Bailey J.A. (1983): Biological perspectives of host-pathogen interactions. In: Bailey J.A., Deverall B.J. (eds.): The Dynamics of Host Defence. Academic Press Australia, Sidney, New York, London, Paris, San Diego, San Francisco, Sao Paulo, Tokyo, Toronto. 1–32

Boller T. (1995): Chemoreception of microbial signals in plant cells. Annu.Rev.Plant Physiol.Plant Mol.Biol. **46**: 189–214

Crute I.R., de Wit P.J.G.M., Wade M. (1985): Mechanisms by which genetically controlled resistance and virulence influence host colonization by fungal and bacterial parasites. In: Fraser R.S.S. (ed.): Mechanisms of Resistance to Plant Diseases. Martinus Nijhoff/Dr.W.Junk, Dordrecht, Boston, Lancaster. 197–309

De Wit P.J.G.M., van den Ackerveken G.F.J.M., Joosten M.H.A.J., van Kan J.A.L. (1989): Apoplastic proteins involved in communication between tomato and the fungal pathogen *Cladosporium fulvum*. In: B.J.J. Lugtenberg (ed.): Signal Molecules in Plant and Plant-Microbe Interactions. Springer-Verlag, Berlin, Heidelberg. 273–280

Ebel J., Cosio E.G. (1994): Elicitors of plant defense responses. Int.Rev.Cytology **148**: 1–36

Heath M.C. (1991): Evolution of resistance to fungal parasitism in natural ecosystems. New Phytol. **119**: 331–343

Keen N.T. (1992): The molecular biology of disease resistance. Plant Mol.Biol. **19**: 109–122

Keen N.T. (1993): An overview of active disease defense in plants. In: Fritig B., Legrand M. (eds.): Mechanisms of Plant Defense Responses. Kluwer Academic Publishers, The Netherlands. 3–11

Knogge W. (1991): Plant resistance genes for fungal pathogens – physiological models and identification in cereal crops. Z.Naturforsch. **46c**: 969–981

Kombrink E., Somssich I.E. (1995): Defense response of plants to pathogens. Adv.Bot.Res. **21**: 1–34

Osbourn A.E. (1996): Preformed Antimicrobial Compounds and Plant Defense against Fungal Attack. Plant Cell **8**: 1821–1831

Osbourn A. (1996): Saponins and plant defense – a soap story. Trends Plant Science **1**: 4–9

Ride J.P. (1985): Non-host resistance in fungi. In: Fraser R.S.S. (ed.): Mechanisms of Resistance to Plant Diseases. Martinus Nijhoff/Dr.W.Junk, Dordrecht, Boston, Lancaster. 29–61

Smart M.G. (1991): The plant cell wall as a barrier to fungal invasion. In: Cole G.T., Hoch H.C. (eds.): The Fungal Spore and Disease Initiation in Plants and Animals. Plenum Press, New York and London. 47–66

Stoessel A. (1983): Secondary plant metabolites in preinfectional and postinfectional resistance. In: Bailey J.A., Deverall B.J. (eds.): The Dynamics of Host Defence. Academic Press Australia, Sidney, New York, London, Paris, San Diego, San Francisco, Sao Paulo, Tokyo, Toronto. 71–122

Wubben J.P., Boller T., Honée G., de Wit P.J.G.M. (1997): Phytoalexins. In: Hartleb H., Heitefuss R., Hoppe H.-H. (eds.): Resistance of Crop Plants against Fungi. Gustav Fischer, Jena, Stuttgart, Lübeck, Ulm. 202–237

Yu L.M. (1995): Elicitins from Phytophthora and basic resistance in tobacco. Proc.Natl. Acad.Sci.USA 92: 4088–4094

Relevant papers

Blein J.P., Milat M.L., Ricci P. (1991): Responses of cultured tobacco cells to cryptogein, a proteinaceous elicitor from Phytophthora cryptogea – possible plasmalemma involvement. Plant Physiol. 95: 486–491

Bostock R.M., Laine R.A., Kuc J.A. (1982): Factors affecting the elicitation of sesquiterpenoid phytoalexin accumulation by eicosapentaeonic and arachidonic acids in potato. Plant Physiol. 70: 1417–1424

Bushnell W.R., Rowell J.B. (1981): Suppressors of defense reactions: A model for roles in specificity. Phytopathology 71: 1012–1014

Davis R.D., Lyon G.D., Darvill A.D., Albersheim P. (1984): Host-pathogen interactions XXV. Endopolygalacturonic acid lyase from Erwinia carotovora elicits phytoalexin accumulation by releasing plant cell wall fragments. Plant Physiol. 74: 52–60

Doke N., Garas N.A., Kuc J. (1980): Effect of host hypersensitivity of suppressors released during germination of Phytophthora infestans cytospores. Phytopathology 70: 35–39

Ham K.-S., Kauffmann S., Albersheim P., Darvill A.G. (1991): Host pathogen interactions XXXIX. A soybean pathogenesis-related protein with β-1,3-glucanase activity releases phytoalexin elicitor-active heat-stable fragments from fungal walls. Molec.Plant-Microbe Interact. 4: 545–552

Kamoun S., Klucher K.M., Coffey M.D., Tyler B.M. (1993a): A gene encoding a host-specific elicitor protein of Phytophthora parasitica. Molecular Plant – Microbe Interactions 6: 573–581

Kamoun S., van West P., de Jong A.J., de Groot K.E., Vleeshouwers V.G.A.A., Govers F. (1997): A gene encoding a protein elicitor of Phytophthora infestans is down-regulated during infection of potato. Molecular Plant – Microbe Interactions 10: 13–20

Kamoun S., Young M., Glascock C.B., Tyler B.M. (1993b): Extracellular protein elicitors from Phytophthora: Host-specificity and induction of resistance to bacterial and fungal phytopathogens. Molec.Plant-Microbe Interact. 6: 15–25

Oku H., Shiraishi T., Kim H.M., Kato T., Saitoh K., Tahara M. (1994): Host selective suppressor of the defense response from Mycosphaerella pinodes. In: Kohomoto K., Yoder O.C. (eds.): Host Specific Toxin: Biosynthesis, Receptor and Molecular Biology. Faculty of Agriculture, Tottori University, Sogo Printing and Publishing Co., Ltd., Tottori, Japan. 49–58

Ricci P., Bonnet P., Huet J.-C., Sallantin M., Beauvais-Cante F., Bruneteau M., Billard V., Michel G., Pernollet J.-C. (1989): Structure and activity of proteins from pathogenic fungi Phythphthora eliciting necrosis and acquired resistance in tobacco. European Journal of Biochemistry 183: 555–563

Ricci P., Trentin F., Bonnet P., Venard P., Mouton-Perronnet F., Bruneteau M. (1992): Differential production of parasiticein, an elicitor of necrosis and resistance in tobacco, by isolates of Phytophthora parasitica. Plant Pathology 41: 298–307

Yoshikawa M., Sugimoto K. (1993): A fungal suppressor of phytoalexin production competes for the elicitor-receptor binding. Naturwissenschaften 80: 374–376

6.1
The Origins of Basic Resistance

Several plant defense mechanisms expressed in active basic resistance against attacking pathogens may also be triggered abiotically. This observation contradicts the plausible assumption that different mechanisms of basic resistance might have been selected during evolution as a response to pathogen attack. Since higher plants must have evolved in the presence of phytopathogens it seems likely that they possessed mechanisms and structures which could be adapted for pathogen defense. Accordingly, it is believed that most, if not all mechanisms involved in basic resistance developed very early in evolution and were secondarily adapted for pathogen defense. Modifications of cell walls by insertion or apposition of lignins, hydroxyproline-rich glycoproteins, or callose could be derived from metabolic pathways selected early in evolution for normal plant growth. For instance, hydroxyproline-rich glycoproteins serve to strengthen cell walls against internal mechanical stress (Corbin et al. 1987), and callose seals plant cell walls and membranes after wounding. Also selection may have occurred for other characters useful in pathogen defense when early plants moved from water to the land, where they had to adapt to a completely new environment. Thus a waxy surface limiting water loss on land may also have repelled some pathogens.

Different levels of disease damage caused by the same pathogen on different plant species may be the result of differences in the effectiveness of the plant's basic resistance. These dissimilar resistances are likely to have been selected for during evolution by many different factors acting on the plant, but probably not solely by the pathogen concerned.

Active basic resistance and race-specific resistance are released by different elicitors acting independently of each other. It seems most likely that the receptors of elicitors of race-specific resistance evolved independently in circumstances where elicitation of active basic resistance did not occur or was suppressed.

The linking of stress reactions with certain elicitors of active basic resistance likely occurred early in plant evolution. We may note that many of the genes induced in this way belong to gene families whose individual members may respond to distinctly different signals. In this way a gene from the family coding for hydroxyproline-rich glycoprotein species is expressed in response to either mechanical wounding or fungal infection (Corbin et al. 1987).

New perspectives have also emerged on the evolution of signal systems which turn on plant gene expression. The signal system for stress-induced phytoalexin synthesis seems to have arisen at the beginning of angiosperm evolution, since it is present in the same form in many angiosperms, mainly in dicots, but less so in monocots. However, during the evolution of different taxa, the universal signal system for phytoalexin synthesis was assigned to different biosynthetic pathways, so that in different plant species it responds to different elicitors. These induction events, which seem now to be understood at least in principle, exhibit great diversity in their combinations in different plant species.

Although plants generally recognize and respond to products derived from fungi, the development of signal systems does not mean that the recognition systems were selected exclusively, or even primarily, by interactions with ancestral fungi. For example, chitin recognition could have evolved from interactions with quite different organisms such as arthropods with chitin exoskeletons. Since plants recognize endogenous elicitors liberated after wounding this may have allowed them also to recognize fungi producing the same or similar substances.

During evolution the basic resistance of plants, like any of their biological properties, will have undergone mutational changes. If a fungal species, by evolving new pathogenicity genes, succeeded in overcoming the basic resistance of a particular plant species, the now susceptible plant might evolve a new, completely different, mechanism of basic resistance that is also effective against this particular pathogen. However, alterations from basic resistance to basic compatibility, and the reverse, require very long periods of time, since the evolution of new pathogenicity genes in the parasite and the acquisition of a new mechanism of basic resistance in the plant requires many independent mutational steps.

Reviews

Heath M.C. (1991): Evolution of resistance to fungal parasitism in natural ecosystems. New Phytol. 119: 331 – 343

Relevant papers

Corbin D.R., Sauer N., Lamb C.J. (1987): Differential regulation of a hydroxyproline-rich glycoprotein gene family in wounded and infected plants. Mol.Cell.Biol. 7: 4337 – 4344

7 Basic Compatibility: The Colonization of Host Plants by Pathogenic Fungi

Basic compatibility between a plant and a pathogen implies the absence of basic resistance against parasitism so that the plant serves as a host. How does basic compatibility arise? What functions are required from each of the partners? According to our present knowledge, four possibilities may be imagined, three of them have been proved to exist in certain plant pathogen systems, whereas one is still hypothetical:

(1) The pathogen is insensitive to the plant's defense reactions. For example, the pathogen could either tolerate high concentrations of defense compounds such as saponins because of its altered membrane structure, or the pathogen detoxifies the saponins by glycosylhydrolases acting as a pathogenicity factor (Bowyer et al. 1995).

(2) The plant does not express defense reactions because the attacking pathogen does not trigger in the plant the release of any defense reaction.

(3) The pathogen suppresses the plant's defense reaction. We have already considered the β-1,3 – and β-1,6 glucans produced by *Phytophthora infestans*, which act as specific suppressors preventing expression of hypersensitive defense reactions in potato slices (Doke et al. 1980) and the supprescins of *Mycosphaerella pinodes*, which prevent the triggering of active basic resistance defense reactions (Oku et al. 1994, see also Chap. 6). Such suppression could involve only one particular defense reaction, a few, or all the defense reactions of the plant.

(4) The pathogen alters the plant's metabolism so that defense reactions cannot be expressed. This has been called "induced susceptibility". In this case, the plant-pathogen interaction does not lead to expression of defense reactions; instead the pathogen triggers reactions that promote parasitism of the plant. However, the reactions must be sufficiently intense and rapid that the expression of defense reactions is overwhelmed. Such a mechanism would be possible only after infection by biotrophic pathogens, because they do not immediately kill their host cells and are able to release metabolites to establish basic compatibility. There is as yet no experimental proof for such a triggering mechanism for basic compatibility, neither is there counter-evidence to exclude it.

However basic compatibility is achieved the pathogen has to express pathogenicity genes tuned to the parasitism of that particular host. Thus, various pathogenicity genes that differ in their mode of action and specificity will be called into play depending, first, on the parasitism strategy used by the pathogen, necrotrophy or biotrophy, and second, on the way it overcomes passive and/or active basic resistance (see Chap. 6). Pathogenicity genes can also be active while colonization is underway, as in the action of genes directing synthesis of phytohormones, which induce formation of tumor-like malformations in plants. However, for the majority of such genes their function and biochemical mechanisms are still unknown. This is due to our lack of knowledge of the individual steps involved in colonization and the genes that control them. In the few plant/pathogen systems where successful identification of particular pathogenicity functions was attempted, the starting points for the analysis were quite different. Two examples will be presented below. In general, most of the activities needed to establish and maintain pathogenicity depend on the concerted action of several genes, some examples are given in Table 3.

First of all, however, an important fact in defining basic compatibility should be addressed. The condition of basic compatibility between host plant and homologous pathogen is still present even if the host plant has mutated to host or cultivar-specific resistance. This latter host plant resistance should be clearly distinguished from the basic resistance or basic incompatibility of non-host plants. The connections between host plant susceptibility and resistance, i.e., under conditions of basic compatibility, are shown as a summary outline in *Fig. 2*. The phenomenon of pathogen defense by host plants triggered by recognition of the pathogen by a race-specific resistant plant as shown in the lower part of *Fig. 2* will be dealt with in Chap. 9, Sects. 9.1 to 9.1.8.

For a number of phytopathogens one of the first requirements for plant colonization is a means of piercing the epidermal cell wall of a leaf to reach the leaf's interior. The first passive barrier is the strong hydrophobic cuticle of the epider-

Table 3. Examples of the concerted activities of pathogenicity genes of phytopathogenic fungi when attacking and colonizing their host plants

Penetration and *exploitation* of the plant, for example, by formation of appressoria, penetration of the host plant by enzymatic and/or mechanical means, formation of haustoria

Detoxification of plant defense substances such as saponins (constitutively synthesized by the plant), phytoalexins (inductively synthesized by the plant after infection)

Synthesis of phytotoxins killing the plant, including non-host selective toxins (NHS-toxin), host-selective toxins (HS-toxin)

Synthesis of plant hormones (phytohormones) causing malformations in the infected plants, for example tumor formation in maize by *Ustilago maydis*, formation of "curly leaf" by the action of *Taphrina deformans* on peaches

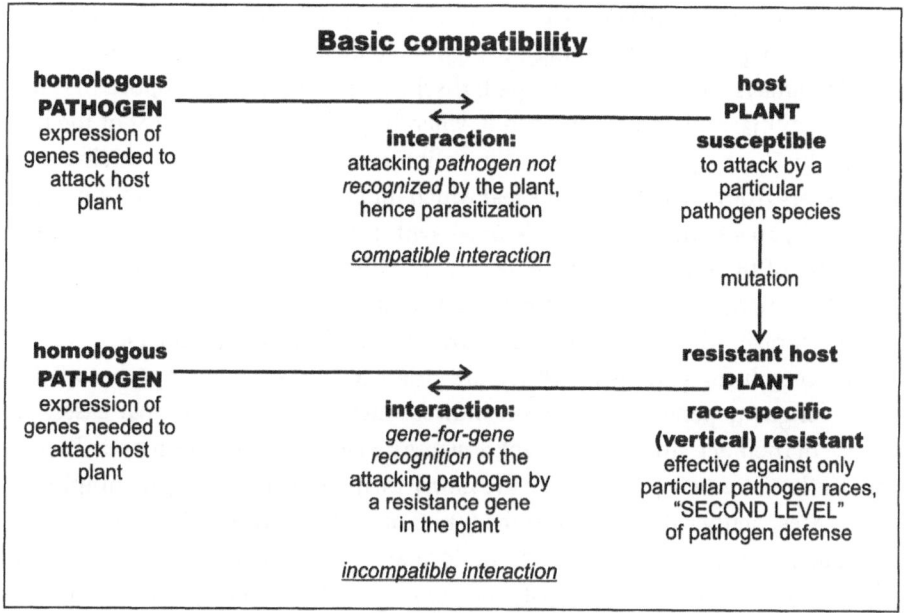

Figure 2. Basic compatibility with attacking pathogens is essential for both susceptibility, resulting in plant parasitism (compatible interaction), and for race-specific, gene-for-gene triggered pathogen resistance (see Sects. 9.1, 9.1.1). Race-non-specific resistance (see Sect. 9.2) depends also on basic compatibility, but not, as in race-specific resistance, on gene-for-gene recognition between host plant and pathogen

mal cell wall. It was supposed for a long time that to pierce the cell wall the pathogen employs both mechanical and enzymatic means (Chap. 3). Investigations by Kolattukudy et al. (Soliday et al.1984, 1989, Woloshuk and Kolattukudy 1986, Dickman et al. 1989) revealed that the cutinase produced by the necrotrophic pathogen *Fusarium solani* f.sp. *pisi* (*Nectria haematococca*) is a pathogenicity factor synthesized by a single gene. The same is true for the cutinases of *Venturia inaequalis*, the apple scab, and *Colletotrichum capsici*, a pathogen on maize and pepper. After growth of *F. solani* and *C. capsici* in a glucose medium, the addition of hydrolyzed cutin induced synthesis of cutinase, secreted into the culture medium, in both fungi. The genes responsible for cutinase synthesis were identified, sequenced and their products characterized in the following way: After isolation of the cutinase mRNAs of both organisms (the mRNA of *F. solani* is about 1050 nucleotides long), corresponding cDNAs were synthesized, cloned and sequenced and used as probes for isolating the corresponding genes from libraries of *F. solani, C. gloeosporioides* and *C. capsici*. Comparison of the amino acid sequences among isolated cutinases revealed sequence homologies in several regions, most probably the domains responsible for cutinase activity.

A nonpathogenic mutant of *F. pisi* was restored to full pathogenicity on uninjured pea leaves after transformation with a cloned cutinase gene. Also, cutinase-negative pathogen mutants were only able to colonize plant leaves with a mechanically injured epidermis. Hence, cutinase fulfills the function of a pathogenicity factor required for breaking through the cuticle of epidermal plant cells. A detailed analysis of the infection path disclosed that a *Fusarium* spore first discharges cutinase, already present in it, when it germinates on the leaf of the host plant. This discharged "sensing" cutinase liberates, by its enzymatic activity, cutin monomers from the cuticle of the leaf which induce cutinase transcription inside the pathogen's germ tube. The *de novo* synthesis of cutinase ultimately causes digestion of the plant's cuticle. However, additional enzymatic activities are required for penetrating the epidermal cell wall, namely pectinase, cellulase and pectinesterase, which are also synthesized by the infecting pathogen.

The search for pathogenicity factors produced by an infecting phytopathogenic fungus depends generally on indirect methods, since, in most cases, the nature of their activities are still unknown. As a rule, after infection one observes the synthesis of new products by the newly formed plant-pathogen complex. In most cases, neither the origin nor the function of the newly synthesized products are known. However, by applying molecular genetic methods it is possible to identify their origin. After identifying, cloning, and sequencing the corresponding genes, it was often possible to deduce clues about their function from their nucleotide sequence. A related experiment is described below.

During compatible interaction of the biotrophic pathogen *Cladosporium fulvum* with tomato plants, after entering the leaves *via* their stomata, the pathogen hyphae spread within the intercellular space without killing the mesophyll cells and without forming microscopically detectable structures serving the pathogen's nutrition, such as appressoria or haustoria. Possibly the pathogen lives in the intercellular space mainly on the sucrose secreted by the mesophyll cells. How is basic compatibility established between the partners and how is it maintained? The working group of de Wit isolated the intercellular fluids from infected leaves. In addition to various dissolved carbohydrates, several proteins with molecular weights of less than 20 kDa were found. These proteins are not synthesized by the pathogen *in vitro*, nor are they found in the intercellular fluids of uninfected leaves. Two of the proteins newly found after infection, ECP1, and ECP2, were thought to be pathogenicity factors synthesized by the parasite (van den Ackerveken et al. 1993, Bazan 1993, Wubben et al. 1994). After purification, the proteins were sequenced and DNA probes prepared to isolate the genes. These *ecp1* and *ecp2* genes were cloned and sequenced. Both genes contained small introns. Gene *ecp1* coded for 96 amino acids, gene *ecp2* for 165. Both polypeptides were processed in two steps, i.e., during secretion by removal of their signal sequences and by action of an extracellular protease, resulting in products of 65 and 142 amino acids, respectively. Both proteins accumulate late in infection in close contact with the host cell walls adjacent to the vascular bundles.

The expression of both genes during the infection process was investigated by using transformants in which the *ecp1* and *ecp2* promoters were fused to the glucuronidase-reporter (GUS-reporter gene) Wild type *C. fulvum* as well as transformants with the *ecp1* and *ecp2* promotors fused to the GUS-reporter gene accumulated similar levels of *ecp1*, *ecp2*, and GUS-transcripts, respectively, as was shown by northern blotting. Enzymatic coupling of glucuronidase with histochemical staining allowed determination of the exact time of gene expression after penetration. It was shown that *ecp1* and *ecp2* are expressed at about 4 days after the hypha entered the interior of the leaf (Wubben et al. 1994). This rather late expression may depend on triggering by host signals. This is in contrast to the expression of the avirulence gene *Avr9* in race-specific resistance (Chap. 9, Sect. 9.1.3.1) immediately after contact between host and pathogen.

The final proof that a gene product functions as a pathogenicity factor requires showing the loss of pathogenicity when the gene in question is defective or missing. Defective pathogens were constructed by gene disruption or by complete deletion of genes *ecp1*, *ecp2* or both. However, they were as pathogenic as the wild type (Marmeisse et al. 1994, Laugé et al. 1997). Possibly, they are not pathogenicity genes, but instead they suppress the synthesis of plant pathogenesis-related (PR) proteins, to promote colonization by the pathogen. However, the absence of these gene functions apparently has no macroscopic effect on the infection symptoms. The nucleotide and amino acid sequences of genes *ecp1* and *ecp2* provided no hint of an enzymatic or structural function for their respective proteins. Hence, definite conclusions about the function of these proteins are still lacking.

Reviews

De Wit P.J.G.M. (1992): Molecular characterization of gene-for-gene systems in plant-fungus interactions and the application of avirulence genes in control of plant pathogens. Annu.Rev.Phytopathol. **30**: 391–418

De Wit P.J.G.M., Joosten M.A.H.J., Honée G., Wubben J.P., van den Ackerveken G.F.J.M., van den Broek H.W.J. (1994): Molecular communication between host plant and the fungal tomato pathogen *Cladosporium fulvum*. Antonie van Leeuwenhoek **65**: 257–262

Heath M.C. (1982): The absence of active defence mechanisms in compatible host-pathogen interactions. In: Wood R.K.S. (ed.): Active Defence Mechanisms in Plants. Plenum Press, New York, London. 143–156

Keen N.T. (1982): Mechanisms conferring specific recognition in gene-for-gene plant parasite systems. In: Wood R.K.S. (ed.): Active Defence Mechanisms in Plants. Plenum Press, New York, London. 67–84

Keen N.T., Staskawicz B. (1988): Host range determinants in plant pathogens and symbionts. Annu.Rev.Microbiol. **42**: 421–440

Knogge W. (1998): Fungal pathogenicity. Curr. Opinion Plant Biol. **1**: 324–328

Köller W. (1991): The plant cuticle. A barrier to be overcome by fungal plant pathogens. In: Cole G.T., Hoch H.C. (eds.): The Fungal Spore and Disease Initiation in Plants and Animals. Plenum Press, New York, London. 219–246

Kolattukudy P.E. (1985): Enzymatic penetration of the plant cuticle by fungal pathogens. Annu.Rev.Phytopathol. **23**: 223–250

Kolattukudy P.E., Podila G.K., Sherf B.A., Bajar M.A., Mohan R. (1991): Mutual triggering of gene expression in plant-fungus interactions. In: Hennecke H., Verma D.P.S. (eds.): Advances in Molecular Genetics of Plant-Microbe Interactions. Dordrecht, Kluwer Academic Publishers, Netherlands. 242–249

Kolattukudy P.E., Podila K., Roberts E., Dickman M.D. (1989): Gene expression resulting from the early signals in plant-fungus interaction. In: Staskawicz B., Ahlquist P., Yoder O. (eds.): Molecular Biology of Plant-Pathogen Interactions. Alan R. Liss, New York. 87–102

Osbourn A. (1996): Saponins and plant defense – a soap story. Trends Plant Science 1: 4–9

Scholtens-Thoma I.M.J., Joosten M.H.A.J., de Wit P.J.G.M. (1991): Appearance of pathogen-related proteins in plant hosts – Relationships between compatible and incompatible interactions. In: Cole G.T., Hoch H.C. (eds.): The Fungal Spore and Disease Initiation in Plants and Animals. Plenum Press, New York, London. 247–265

Vanderplank J.E. (1986): Specific susceptibility and specific feeding in gene-for-gene systems. Advances in Plant Pathology 5: 199–223

Van Etten H.D. (1979): Relationship between tolerance to isoflavonoid phytoalexins and pathogenicity. In: Daly J.M., Uritani I. (eds.): Recognition and Specificity in Plant Host-Parasite Interactions. University Park Press, Baltimore. 301–316

Wubben J.P., Boller T., Honée G., De Wit P.J.G.M. (1997): Phytoalexins. In: Hartleb H., Heitefuss R., Hoppe H.-H. (eds.): Resistance of Crop Plants against Fungi. Gustav Fischer, Jena, Stuttgart, Lübeck, Ulm. 202–237

Relevant papers

Bazan J.F. (1993): Emerging families of cytokines and receptors. Curr.Biol. 3: 603–606

Bowyer P., Clarke B.R., Lunness P., Daniels M.J., Osbourn A.E. (1995): Host range of a plant pathogenic fungus determined by a saponin detoxifying enzyme. Science 267: 371–374

Dickman M.B., Podila G.K., Kolattukudy P.E. (1989): Insertion of cutinase gene into a wound pathogen enables it to infect intact host. Nature 342: 446–448

Doke N., Garas N.A., Kuc J. (1980): Effect on host hypersensitivity of suppressors released during germination of *Phytophthora infestans* cytospores. Phytopathology 70: 35–39

Laugé R., Joosten M.H.A.J., Van den Ackerveken G.F.J.M., Van den Broek H.W.J., De Wit P.J.G.M. (1997): The in planta produced extracellular proteins ECP1 and ECP2 of *Cladosporium fulvum* are virulence factors. Molecular Plant – Microbe Interactions 10: 725–734

Marmeisse R., Van den Ackerveken G.F.J.M., Goosen T., De Wit P.J.G.M., Van den Broek H.W.J. (1994): The *in-planta* induced *ecp2* gene of the tomato pathogen *Cladosporium fulvum* is not essential for pathogenicity. Curr.Genet. 26: 245–250

Oku H., Shiraishi T., Kim H.M., Kato T., Saitoh K., Tahara M. (1994): Host selective suppressor of the defense response from *Mycosphaerella pinoides*. In: Kohomoto K., Yoder O.C. (eds.): Host Specific Toxin: Biosynthesis, Receptor and Molecular Biology. Faculty of Agriculture, Tottori University, Sogo Printing and Publishing Co., Ltd., Tottori, Japan. 49–58

Soliday C.L., Dickman B., Kolattukudy P.E. (1989): Structure of the cutinase gene and detection of promotor activity in the 5'-flanking region by fungal transformation. J.Bacteriol. 171: 1942–1951

Soliday C.L., Flurkey W.H., Okita T.W., Kolattukudy P.E. (1984): Cloning and structure determination of cDNA for cutinase, an enzyme involved in fungal penetration of plants. Proc.Natl.Acad.Sci.USA 81: 3939–3943

Van den Ackerveken G.F.J.M., Van Kan J.A.L., Joosten M.H.A.J., Muisers J.M., Verbakel H.M., De Wit P.J.G.M. (1993): Characterization of two putative pathogenicity genes in the fungal tomato pathogen *Cladosporium fulvum*. Molec.Plant-Microbe Interact. 6: 210–215

Woloshuk C.P., Kolattukudy P.E. (1986): Mechanism by which contact with plant cuticle triggers cutinase gene expression in the spores of *Fusarium solani* f.sp.*pisi*. Proc.Natl. Acad.Sci.USA 83: 1704–1708

Wubben J.P., Joosten M.H.A.J., De Wit P.J.G.M. (1994): Expression and localization of two *in planta* induced extracellular proteins of the fungal tomato pathogen *Cladosporium fulvum*. Molec.Plant-Microbe Interact. 7: 516–524

8 Phytotoxins: The Weapons of Necrotrophic Phytopathogenic Fungi

Symptoms of plant disease caused by necrotrophic fungi, such as chlorosis (bleaching) or wilting of leaves, necroses, rotting of either parts or of the whole plant, may also be found after treatment of the plants with the sterile, filtered culture fluids of these fungi. Evidently, these pathogens synthesize substances causing disease symptoms in plants that are synthesized not only *in planta*, but also during growth in artificial medium. If they are still active at high dilution, they are called toxins. Most toxins have low molecular weights, up to about 1000 Daltons and are also toxic to non-host plants. Some of the toxins are poisonous for animals and humans, e.g., toxins synthesized by fungi of the genus *Fusarium*. However, within the same fungal genus toxins may be produced that are barely toxic for plants but which are poisonous for animals and humans. Toxins produced by phytopathogenic fungi and poisonous to plants are called phytotoxins irrespective of their action on other organisms.

Phytotoxins can spread in plants by diffusion from the site of fungal infection to adjacent tissue or through transport via the plant's apoplast or xylem. Phytotoxins may function as virulence factors, i.e., they act in a quantitative manner intensifying the plant's disease symptoms, or they may act like pathogenicity factors, i.e., qualitatively, and hence are responsible exclusively for development of disease symptoms.

Two classes of phytotoxins can be distinguished: The majority belong to the non-selective or *non-host*-specific toxins (NHS-toxins). These are poisonous to all plants, not only the plant serving as host for the toxin-producing pathogen. Hitherto, no plants have been found that are insensitive to the action of isolated NHS toxins, whereas mutant host plants have been found that are resistant to colonization by a pathogen producing a toxin. Also, pathogen mutants defective in toxin synthesis have been isolated which have thereby lost their pathogenicity. Phytopathogenic fungi producing non-host-selective toxins are among the most injurious organisms in agriculture, and cause substantial economic losses.

The so-called host-selective or host-specific toxins (HS-toxins), belong to the second and smaller class of phytotoxins. They are active only on host plants carrying genetically determined sensitivity for the particular toxin.

All NHS-toxins are virulence factors, hence they intensify the severity of disease symptoms in the plant. By contrast, most of the HS-toxins are pathogenicity factors, i.e., they are exclusively responsible for the appearance of disease symptoms.

Phytotoxins of both classes are effective in many different ways, as one might expect from the various disease symptoms observed with different toxins. In most cases toxins act more or less directly, quickly causing death of the infected plant cell. The process of killing has been shown experimentally to proceed in a few distinct steps. However, toxins may also act indirectly via activation of metabolic pathways which result in the synthesis of phytoalexins. In this case, the toxin acts as an elicitor of a hypersensitive reaction. Toxins may also suppress the expression of induced resistance in the infected plant (see Chap. 10). In the following three sections NHS- and HS-toxins will be treated in more detail together with the genetic determination of the biosynthesis of HS toxins and the nature of mutations to resistance against them.

Reviews

Durbin R.D. (1981): Toxins in Plant Disease. Academic Press, New York, London, Toronto, Sidney, San Francisco

Durbin R.D. (1983): The biochemistry of fungal and bacterial toxins and their modes of action. In: Callow J.A. (ed.): Biochemical Plant Pathology. John Wiley & Sons, Chichester. 137–162

Graniti A. (1991): Phytotoxins and their involvement in plant diseases. Introduction. Experientia 47: 751–755

Graniti A., Durbin R.D., Ballio A. (1989): Phytotoxins and Plant Pathogenesis. NATO ASI Series Cell Biology. Springer-Verlag, New York, London, Paris, Tokyo

Hoppe H.-H. (1997): Fungal phytotoxins. In: Hartleb H., Heitefuss R., Hoppe H.-H. (eds.): Resistance of Crop Plants against Fungi. Gustav Fischer, Jena, Stuttgart, Lübeck, Ulm. 58–83

Scheffer R.P. (1991): Role of toxins in evolution and ecology of plant pathogenic fungi. Experientia 47: 804–811

Sequeira L. (1983): Recognition and specificity between plants and pathogens. In: Kommendahl T., Williams P.M. (eds.): Challenging Problems in Plant Health. Am. Phytopathol. Soc., St. Paul. 301–310

Shaner G., Stromberg E.L., Lacy G.H., Barker K.R., Pirone T.P. (1992): Nomenclature and concepts of pathogenicity and virulence. Annu.Rev.Phytopathol. 30: 47–66

Yoder O.C. (1980): Toxins in pathogenesis. Annu.Rev.Phytopathol. 18: 103–129

8.1
Non-Host-Selective Toxins

The chemical structure of many non-host-selective toxins (NHS-toxin) has already been determined. However, corresponding investigations of the basis of their physiological and biochemical actions in plants are still lacking. In contrast, there has been more research work on the effects of non-host-selective phytotoxins on animal and human cells or corresponding tissues. This research was prompted by the occurrence of serious symptoms of poisoning after feeding animals with seeds, tubers, or roots diseased by fungal pathogens, or when such materials were eaten by humans.

Some of the most thoroughly investigated examples of NHS toxins include: brefeldin A, produced by *Penicillium decumbens* and *Alternaria carthami*; cercosporin, from *Cercospora kikuchii* and other species of *Cercospora*; fusicoccin, from *Fusicoccum amygdali*; ophiobolin, from *Cochliobolus miyabeanus*; tentoxin, from *Alternaria alternata*; zinniol, produced by different species of *Alternaria* and *Phoma macdonaldii*; and fusaric acid, synthesized by different species of *Fusarium*. All these toxins are products of the secondary metabolism of their respective fungi. Their targets are primarily host plant organelles including membranes, mitochondria, microsomes or chloroplasts. As mentioned above, the isolated NHS toxins of necrotrophic fungi are poisonous to most other plants, in addition to their own host plant(s). Therefore, the spectrum of action of each toxin is much broader than the host range of the pathogen producing it.

Each of the NHS toxins seems to act on plants by a highly specific mechanism. However, the site and mode of action of very few have been analyzed thus far in any detail. The toxin-binding sites are located in the microsomal and plasma membranes of most land plants, as demonstrated by experiments showing specific binding of ^{3}H-labeled fusicoccin to membranes of seed plants, ferns and some bryophytes. Specific fusicoccin-binding sites are absent in prokaryotes, algae, fungi and animal cells (Meyer et al. 1993). The toxins, fusicoccin and tentoxin, both appear to affect phosphorylation. Fusicoccin activates phosphorylation of the membrane bound H^+-ATPase thereby increasing uptake of K^+ and efflux of H^+ ions whereas tentoxin inhibits energy transfer in chloroplasts during light-dependent phosphorylation. The toxic action of fusicoccin could be due either to its receptor belonging to the widespread 14-3-3 superfamily of eukaryotic proteins, the different functions of which are regulated by their phosphorylation, or to its interference with phosphorylation of individual components of signal transduction chains. The broad spectrum of action of tentoxin on different plant cells could be similarly linked to phosphorylation.

Because of their highly specific reaction mechanisms, some of the NHS phytotoxins were used as tools for studying physiological and biochemical reactions. Thus, fusicoccin was utilized in plants for analyzing transport processes through membranes and tentoxin as a blocker of phosphorylation. Brefeldin A was employed in animals to investigate intracellular transport processes.

Reviews

Ballio A. (1991): Non-Host-Selective Fungal Phytotoxins – biochemical aspects of their mode of action. Experientia **47**: 783–790

Durbin R.D. (1981): Toxins in Plant Disease. Academic Press, New York, London, Toronto, Sidney, San Francisco

Durbin R.D. (1983): The biochemistry of fungal and bacterial toxins and their modes of action. In: Callow J.A. (ed.): Biochemical Plant Pathology. John Wiley & Sons, Chichester. 137–162

Graniti A. (1991): Phytotoxins and their involvement in plant diseases. Introduction. Experientia **47**: 751–755

Graniti A., Durbin R.D., Ballio A. (1989): Phytotoxins and Plant Pathogenesis. NATO ASI Series Cell Biology. Springer-Verlag, New York, London, Paris, Tokyo

Hoppe H.-H. (1997): Fungal phytotoxins. In: Hartleb H., Heitefuss R., Hoppe H.-H. (eds.): Resistance of Crop Plants against Fungi. Gustav Fischer, Jena, Stuttgart, Lübeck, Ulm. 58–83

Knogge W. (1996): Fungal Infection of Plants. Plant Cell **8**: 1711–1722

Mitchell R.E. (1984): The relevance of non-host-specific toxins in the expression of virulence by pathogens. Annu.Rev.Phytopathol. **22**: 215–245

Rudolph K. (1976): Non-specific toxins. In: Heitefuß R., Williams P.H. (eds.): Encyclopedia of Plant Pathology. Springer-Verlag, Berlin, Heidelberg, New York. 270–315

Sequeira L. (1983): Recognition and specificity between plants and pathogens. In: Kommendahl T., Williams P.M. (eds.): Challenging Problems in Plant Health. Am. Phytopathol. Soc., St. Paul. 301–310

Relevant papers

Meyer C., Waldkötter K., Sprenger A., Schlösser U.G., Luther M., Weiler E.W. (1993): Survey of taxonomic and tissue distribution of microsomal binding sites of the non-host selective fungal phytotoxin, Fusicoccin. Z.Naturforsch.C **48c**: 595–602

8.2
Host-Selective Toxins

Interactions between pathogens producing host-selective (HS) toxins and their host plants have been thoroughly investigated only in a few cases, involving pathogens of the genus *Cochliobolus* and *Alternaria*. The interactions follow the principle of a signal-sensor reaction, as observed in pathogen defense mediated by recognition between a homologous pathogen and its host plant (see Chap. 2): The HS-toxin represents the signal recognized by the sensor of the host plant (host recognition). The sensor is a receptor protein that specifically binds the HS-toxin. This binding is classified as "recognition" of the pathogen if it is followed on the plant's side by a rapid "response". Described in general terms, the response corresponds to support of the pathogen in its efforts to colonize the plant. Receptor proteins which recognize a particular HS-toxin do not appear to code for any function essential for the plant, since variants of the very same plant that are not sensitive to this HS-toxin are common and fully viable.

The recognition of an HS-toxin proceeds in three steps: (1) the germination tube of the fungal spore, in contact with the plant's surface, releases toxin molecules that act as signals before it penetrates the cell wall; (2) the HS-toxin signal molecules bind to the corresponding plant receptors; and (3) the reaction triggered by this specific signal-sensor binding is transduced, generating additional signal molecules along signal transduction chains. This amplification paralyzes the plant's defense mechanisms prior to cell killing. Thus plant cells are converted by the pathogen into an easily colonized state by preventing the plant from expressing functions inhibiting pathogen penetration. Prevention of this penetration inhibition may be independent of the killing activity of the toxin.

An HS-toxin-producing pathogen can still colonize its host plant even when the plant lacks the specific receptor to recognize the HS-toxin. However, under these circumstances no disease symptoms will develop, as neither the pathogen nor its toxin is "recognized". The plant, being non-sensitive or resistant to the HS-toxin (genotype *sens⁻*), develops no, or at most only very weak, disease symptoms. In this case the cultivar's classification as "resistant" characterizes only its insensitivity to the reduction in crop yield that would be expected if the toxin were recognized by the plant. Also no toxin effects are observed if the host plant is sensitive to the toxin (*sens⁺*) but the pathogen is unable to produce active toxin (genotype *tox⁻*). Corresponding *tox⁻* mutants exist, for example, for the pathogen *Cochliobolus victoriae* (see below), and *sens⁻* or resistant mutants for its host plant oat. Both types of mutations, in the pathogen or the plant, result in loss of toxin function and demonstrate that something like two "corresponding factors" are required for HS-toxin action: on the pathogen side, the ability to synthesize an active HS-toxin; on the host plant side, production of the specific receptor for it. Recognition between both gene products corresponds to a signal-sensor reaction.

Crossing experiments revealed that the genetic determinant for each corresponding factor in pathogen and plant maps to a single chromosomal locus suggesting that in the pathogen and the plant only one gene may code for the signal-sensor reaction. In other words, the action of HS-toxins seems to depend on a gene-for-gene interaction comparable to that found for expression of race-specific resistance in correspondingly resistant host plants (see Chap. 9, Sect. 9.1.1). However, there is one difference concerning the outcome of both gene-for-gene interactions. Recognition between an HS-toxin and the susceptible host plant results in killing of the plant, or at least of parts of its tissue, followed by extensive colonization by the pathogen. Therefore, HS-toxins may be considered as agents for establishing compatibility. By contrast, recognition between a pathogen and its race-specific resistant host plant, i.e., the pathogen's avirulence gene product (Chap. 9, Sect. 9.1) and the plant's resistance gene product effecting expression of race-specific resistance, prevents colonization by the fungus, i.e., the attacking pathogen is repelled. (It should be stressed that the avirulence gene product, as elicitor of the hypersensitive reaction – like an HS-toxin – directly kills the infected cell; see hypersensitive cell death or programmed cell death, PCD, in Chap. 5.)

The outcome of interactions between host plants of genotypes *sens⁺* or *sens⁻* and pathogens of genotypes *tox⁺* or *tox⁻* is easy to survey in a so-called quadratic check. Table 4 shows that toxin action, i.e., appearance of disease symptoms, is observed only when *tox⁺* pathogens and *sens⁺* plants interact. The three other pathogen/plant combinations result in no or only very weak disease symptoms. The same specificity pattern is obtained with isolated HS-toxins as shown in Table 4.

As mentioned above, HS-toxins may sometimes act as pathogenicity factors, implying that symptoms develop only if the plant recognizes the toxin. If, how-

Table 4. Quadratic check based on toxin sensitivity

		Host plants	
		Not toxin susceptible (*sens⁻*)	Toxin susceptible (*sens⁺*)
Pathogens	No toxin production (*tox⁻*)	No parasitism or parasitism without disease symptoms	No parasitism or parasitism without disease symptoms
	Toxin production (*tox⁺*)	No parasitism or parasitism without disease symptoms	**Parasitism with killing** of host plant

Results of infecting either toxin sensitive (*sens⁺*) or insensitive (*sens⁻*) host plants with a pathogen producing an HS-toxin (*tox⁺*) or a mutant defective in toxin production (*tox⁻*).

ever, the pathogen is *tox⁻* or the host *sens⁻* or both genes are defective, the plant may still become parasitized – although somewhat more weakly – but no disease symptoms develop. In other cases the infecting pathogens may still cause disease symptoms in plants even if toxin production is defective due to a mutation. Under such circumstances, the toxin would only intensify expression of disease symptoms, and the HS-toxin is classified as a virulence factor.

HS-toxins are substances of very different and unusual chemical structures, as, for example, circular peptides, sesquiterpene galactofuranosides, linear polyketols and linear β-ketoalcohols, epioxydecatrienoinic acids, all with molecular masses below 1000 Daltons; the only exception known so far is a polypeptide of *Pyrenophora tritici-repentis*, the Pts-toxin, of 13.2 kDa. Some HS-toxins are poisonous at concentrations as low as 10^{-9} M. They are formed during secondary metabolism, are coded for either by one chromosomal locus or by gene families, and they are frequently produced as mixtures of different isomers or structurally closely related molecules. Because of their stereochemical properties, they bind specifically to the plasma membrane or membranes of mitochondria or chloroplasts, triggering distinct physiologic reactions at these sites. One of the earliest microscopically observed reactions after binding of an HS-toxin to its receptor is prevention of the plant's defense even before the pathogen starts its penetration. During colonization of the plant's interior by the pathogen, the phytotoxin may be further transported via the phloem or xylem resulting in killing of the entire plant or at least of some of its tissue. A list of HS-toxins known up to 1990 was compiled by Kohmoto and Otani.

HS-toxins are frequently produced by acomycete pathogens in the genera *Alternaria* and *Cochliobolus*. Accordingly, both genera were chosen for investigating the mechanism of action and the physiology of these toxins. Various pathotypes or races of *Alternaria alternata* are specialized in that they cause disease on particular cultivated plants such as pear, apple, strawberry, citrus fruits, tomato or tobacco. *A. alternata* seems to have evolved new HS-toxin variants by mutation. In this way many new, extremely specialized, and highly virulent *forma speciales* have evolved from a saprophytic fungal species that potentially can act also as an opportunistic pathogen.

Comparable observations were reported for three species of the genus *Cochliobolus* which form quite different HS-toxins:

(1) The HMT- or T-toxin, is produced by *C. heterostrophus* (*Helminthosporium maydis*) race T, a pathogen on maize plants. The T-toxin is a linear polyketol toxic only to maize cultivars carrying the cytoplasmically inherited gene *tms*, for pollen sterility (male sterile cytoplasm). This gene *T-urf13* codes for a receptor specific for T-toxin recognition. *Tox⁻* strains unable to produce active toxin still cause disease symptoms, albeit much weaker, on maize. Therefore, the T-toxin was classified as a virulence factor.

(2) The HC-toxin of *C. carbonum*, race 1, also a pathogen of maize, is a cyclic tetrapeptide which is detoxified in most maize cultivars. During breeding for crop improvement, a gene called *Hm*, responsible for detoxification but at the time of these breeding experiments not yet discovered, was accidentally deleted. Whereas after infection with *C. carbonum* all other maize cultivars develop only mild symptoms, this newly bred cultivar exhibited an extremely high susceptibility to this infection, resulting in devastating economic damages in the USA. This situation led to the discovery of the HC-toxin and the subsequent, very thorough investigation of its effects. HC-toxin acts like a virulence factor, i.e., it intensifies disease symptoms.

(3) The HV-toxin, also called victorin C, of the oat pathogen *C. victoriae* is an acyclic combination of glyoxylic acid with five unusual amino acids. It is toxic only for oat cultivars carrying the chromosomal marker *Vb*. In contrast to the foregoing example, after infection of *Vb* cultivars with *tox⁻* strains, no disease symptoms appear, and the same holds true for infection of cultivars lacking the marker *Vb* with *tox⁺* pathogens. Thus, victorin acts like a pathogenicity factor, it is capable alone of causing the observed disease symptoms. The *Vb* locus seems to be identical to the gene *Pc-2* of oat plants conferring race-specific resistance against the biotrophic pathogen *Puccinia coronata* (Rines and Luke 1985). Therefore the locus is also called *Vb/Pc-2*. Apparently, there is a close functional relationship between the action of victorin on this toxin-susceptible oat cultivar and the action of the avirulence factor of *P. coronata* triggering expression of race-specific resistance of this resistant cultivar (see also Chap. 9, Sect. 9.1.3.3).

The fact that HS-toxins are associated with pathogens of cultivated plants suggests that selection of the different HS-toxin-producing pathogens might be the result of coevolution of necrotrophic pathogens and their host plants caused by human activities. The cultivation of genetically homogeneous crops in monocultures has probably strongly selected necrotrophic pathogens able to produce highly poisonous HS-toxins. One particularly instructive example was the rapid spread of the truly devastating Southern corn leaf blight epidemic in the USA in 1970–71 of the maize pathogen *C. heterostrophus* race T on maize cultivars carrying the genetic marker *tms* (see above (1)).

Reviews

Brettell R.I.S., Pryor A.J. (1986): Molecular approaches to plant and pathogen genes. In: Blonstein A.D., King P.J. (eds.): Plant Gene Research: A Genetic Approach to Plant Biochemistry. Springer-Verlag, Vienna, New York. 233–246

Daly J.M. (1984): The role of recognition in plant disease. Annu.Rev.Phytopathol. 22: 273–307

Ellingboe A.H. (1981): Changing concepts in host-pathogen genetics. Annu.Rev.Phytopathol. 19: 125–143

Graniti A. (1991): Phytotoxins and their involvement in plant diseases. Introduction. Experientia 47: 751–755

Hoppe H.-H. (1997): Fungal phytotoxins. In: Hartleb H., Heitefuss R., Hoppe H.-H. (eds.): Resistance of Crop Plants against Fungi. Gustav Fischer, Jena, Stuttgart, Lübeck, Ulm. 58–83

Kohmoto K., Otani H. (1991): Host recognition by toxigenic plant pathogens. Experientia 47: 755–764

Novacky A. (1991): The plant membrane and its response to disease. In: Cole G.T., Hoch H.C. (eds.): The Fungal Spore and Disease Initiation in Plants and Animals. Plenum Press, New York, London. 363–378

Panaccione D.G. (1993): The fungal genus *Cochliobolus* and toxin-mediated plant disease. Trends Microbiol. 1: 14–20

Schäfer W. (1994): Molecular mechanisms of fungal pathogenicity to plants. Annu.Rev. Phytopathol. 32: 461–477

Scheffer R.P. (1991): Role of toxins in evolution and ecology of plant pathogenic fungi. Experientia 47: 804–811

Sequeira L. (1983): Recognition and specificity between plants and pathogens. In: Kommendahl T., Williams P.M. (eds.): Challenging Problems in Plant Health. Am. Phytopathol. Soc., St. Paul. 301–310

Ullstrup A.J. (1972): The impact of Southern corn leaf blight epidemic of 1970–71. Annu.Rev.Phytopathol. 10: 37–50

Walton J.D. (1996): Host-Selective Toxins: Agents of Compatibility. Plant Cell 8: 1723–1733

Walton J.D., Panaccione D.G. (1993): Host-selective Toxins and disease specificity: Perspectives and progress. Annu.Rev.Phytopathol. 31: 275–303

Yoder O.C. (1980): Toxins in pathogenesis. Annu.Rev.Phytopathol. 18: 103–129

Relevant papers

Rines H., Luke H.H. (1985): Selection and regeneration of toxin-insensitive plants from tissue cultures of oats (*Avena sativa*) susceptible to *Helminthosporium victoriae*. Theor.Appl.Genet. 71: 16–21

8.3
Genetics of Toxin Biosynthesis and Toxin Resistance

Investigating the genetics of phytotoxin formation is one of the most powerful tools for elucidating its biosynthesis. Experiments with host-selective toxins produced by different species of the genus *Cochliobolus* have been especially informative. Analysis of the nature of toxin-resistant plant mutants has contributed in a similar way to an understanding of the mechanism of phytotoxin actions. Similar experiments with other genera of toxin producing necrotrophic fungal pathogens were frustrated by the fact that many of these necrotrophs lack a sexual cycle, e.g., all species of the genus *Alternaria* as well as most of those of the genus *Fusarium*. Since neither of these genera can be sexually crossed, only parasexual or molecular genetic methods could allow identification of genes involved in the production of phytotoxins.

Analysis of the genetic basis of HS-toxin synthesis in the genus *Cochliobolus* posed two questions: (1) How many genes participate in the formation of HS-toxins? (2) Are different genes responsible for expression of virulence and toxin synthesis, or can both properties be attributed to the same gene? When tox^- and tox^+ genotypes of *Cochliobolus heterostrophus* and of *C. carbonum* were crossed, a 1 : 1 segregation for the tox^+ and tox^- phenotypes was observed in the progeny. This result showed that toxin production is coded by a single chromosomal locus. However, the result leaves unresolved the question of whether a single or several closely linked genes, not separated in this crossing experiment, are responsible for toxin synthesis. Regarding the second question, namely, whether both phenotypes, virulence and toxin synthesis, are determined by the same gene, it was found that all tox^+ strains of *Cochliobolus*, i.e., those strains carrying the gene locus *TOX1* for synthesis of HMS-toxin in infection experiments, are both highly virulent and produce high levels of toxin. Cosegregation of the two phenotypes suggests that they are determined by the same locus. Furthermore, the phenotype of toxin production seems to be inherited dominantly as shown by experiments with heterokaryons forced by employing complementing auxotrophic markers, and the amount of toxin production is regulated genetically. The loci *TOX2* and *TOX3*, coding for HC-toxin and HV-toxin of *C. carbonum* and *C. victoriae*, respectively, segregated independently in crosses between the species. This demonstrates that *TOX* loci of both species map to different sites, probably on different chromosomes.

The three gene loci *TOX1*, *TOX2*, and *TOX3*, coding for the toxins HMT-toxin, HC-toxin, and victorin of *C. heterostrophus, C. carbonum, and C. victoriae*, respectively, are coded by gene families, that segregate in crosses as single chromosomal loci. Thus victorin C is the main component of the toxin victorin coded by the *TOX3* gene family. Each of the three gene families exhibits a gene-for-gene relationship with a corresponding gene in the host plant coding for the receptor protein that specifically recognizes the particular fungal toxin. For example, the susceptibility of maize plants to HMT-toxin or T-toxin of *C. heterostrophus* race T (synonym: *Bipolaris maydis*) and the PM-toxin of *Mycosphaerella zea-maydis* (synonym: *Phyllosticta maydis*) depends on the presence of the particular genetic marker in the plant, coded for by the gene *T-urf13* and expressed as the protein T-URF13. The latter protein was shown to specifically bind ^{3}H-labeled toxin and hence is considered to be the site of action, i.e., the receptor, for the HMT- and PM-toxins (Dewey et al. 1986, 1988). The plant's T-URF13 product also functions as a so-called male sterility factor. Male sterility refers to the inability of maize plants to produce functional pollen, a property which depends on the presence of a 13 kDa protein in the inner mitochondrial membrane. As shown by DNA sequencing, its coding gene, *T-urf13* arose from a single recombination event within the mitochondrial DNA, that fused the promoter of ATPase subunit 6 with the gene for ribosomal 265 RNA. This fusion created the new gene *T-urf13*, the product of which causes pollen sterility. However, Gene *T-urf13* is inherited as a cytoplasmic determinant since it is a mitochondrial trait that is only transmitted through the mother plant. Since the plant's toxin susceptibility depends on the presence of this particular protein, toxin susceptibility is inherited as a dominant trait, whereas resistance, i.e., absence of this protein and hence insensitivity to the toxin, is recessive.

With the HC-toxin of *C. carbonum*, the situation is reversed, since toxin susceptibility of the host maize plant is a recessive trait and resistance is dominant. The resistance results from the activity of an HC-toxin reductase, dihydroflavonol-4-reductase, which is coded for by the gene *Hm*, which detoxifies HC-toxin. Clearly, lack of this gene and the corresponding enzyme activity manifests as a recessive character. HC-toxin reductase, or its functional equivalent, is present in all genotypes of maize as well as other in grasses. Furthermore, gene *Hm* shows some homology to a similar activity in petunia, snapdragon (*Antirrhinum*), and some grasses (*Gramineae*). As mentioned already, the gene *Hm* was deleted accidentally when breeding for a particular maize cultivar. After its release the cultivar turned out to be highly susceptible to infection by *C. carbonum*, a fungus causing little damage on other maize cultivars. Investigation of the high sensitivity of the newly bred cultivar led to discovery of the host selective-acting HC-toxin produced by *C. carbonum* and the inability of the particular cultivar to detoxify it because of the lack of the dihydroflavonol reductase coded for by gene *Hm* (Johal et al. 1992). The cloned *Hm* gene was the first well characterized plant resistance gene, the function of which became fully understood. In contrast to the HMT-toxin, the receptor for the HC-toxin in maize remains to be identified.

Oat cultivars carry a genetic locus, *Vb*, which renders them susceptible to victorin C produced by the pathogen *C. victoriae*. Two proteins of 100 kDa and 15 kDa were identified in oat cultivars that specifically bind this toxin. In victorin-resistant cultivars, only the 15 kDa protein binds victorin C, whereas in the toxin-susceptible cultivar, carrying the *Vb* marker, both, 15 kDa and 100 kDa proteins bind victorin. This suggests that the 100 kDa protein of the susceptible cultivar is the binding site responsible for toxin sensitivity. The 100 kDa and 15 kDa proteins (also called P- and H-proteins) are localized within the mitochondria of oat plants where they form, together with two other proteins, L and T, the multienzyme complex of glycine decarboxylase which *in vivo* is composed of 42 subunits (Wolpert and Macko 1989, Wolpert et al. 1994, Navarre and Wolpert 1995). The activity of glycine decarboxylase is inhibited *in vivo* and *in vitro* by victorin. Glycine decarboxylase is widespread in nature, involved in the glycine-serine conversion, and is found in prokaryotes and eukaryotes. In green plants glycine decarboxylase plays a crucial role in the photorespiratory cycle and mutations in this cycle are lethal. Consequently, inhibition by victorin of the cycle should be lethal to green tissues. For now, it is still unknown (1) why victorin binds in a ligand- and genotype-specific manner only to the P-protein of glycine decarboxylase of oat plants of genotype *Vb/Pc-2* and (2) what the specific physiological effects are of this binding.- Table 5 summarizes some of the data on the toxins synthesized by the three species of the genus *Cochliobolus*.

None of the host-selective toxins analyzed so far has a chemical structure corresponding to a primary gene product, i.e, a polypeptide, although some of the toxins may contain peptides. Rather, HS-toxins are products of pathogen secondary metabolism, synthesized in several steps that seem to involve pathways that are not involved in the pathogen's energy and reproduction metabolism, with each step of the synthesis determined by a particular gene. This raises an interesting question, namely, how this situation can be reconciled with the notion that

Table 5. Comparison of the host-selective toxins produced by three species of *Cochliobolus*

Pathogen	Toxin	Gene	Host plant	Receptor gene	Reasons for resistance
Cochliobolus heterostrophus (synonym: *Helminthosporium maydis*)	**HMT-toxin** (synonym: T-toxin) Linear polyketol Virulence factor	*TOX1*	Maize	*t-urf13* Dominant	*t-urf13* **Not** present
Cochliobolus carbonum	**HC-toxin** Cyclic tetrapeptide Virulence factor	*TOX2*	Maize	n.i. Recessive	Detoxification by gene *Hm*
Cochliobolus victoriae (synonym: *Helminthosporium victoriae*)	**HV-toxin** Main component victorin C – a derivative of glyoxylic acid- Pathogenicity factor	*TOX3*	Certain oat cultivars	*Vb* Dominant	*Vb* **Not** present

n.i., not identified.

in genetic crosses toxin production segregates as a single locus. To answer this question two possibilities can be envisaged: either the genes participating in toxin synthesis are very closely linked, i.e., they are clustered on one chromosomal locus and cannot be separated by currently available methods in crosses by genetic recombination, or indeed only a single gene codes for toxin synthesis. However, this gene would need to contain several different domains coding for a multifunctional enzyme catalyzing all the different steps required for toxin synthesis. An example of such a complex chromosomal structure is the *TOX2*-locus, coding for the HC-toxin of *C. carbonum*. This very complex locus of 540 kb contains three clustered genes, each present in two or three copies, plus two additional genes as yet only partially identified. Two or more of these appear to code for multifunctional enzymes. – The locus coding synthesis of the HMT-toxin of *C. heterostrophus* is somewhat different. This locus contains at least two genes located on different chromosomes. However, in crosses both genes segregate like very closely linked genes, i.e., as a single locus, since they map immediately to a reciprocal chromosomal translocation point. Thus the biochemical and molecular data indicate that the genetic determination of most examples of HS-toxin sensitivity does not involve strict gene-for-gene relationships determining toxin susceptibility, i.e., it does not depend exclusively on one gene in the host plant and one in the pathogen.

Reviews

Brettell R.I.S., Pryor A.J. (1986): Molecular approaches to plant and pathogen genes. In: Blonstein A.D., King P.J. (eds.): Plant Gene Research: A Genetic Approach to Plant Biochemistry. Springer-Verlag, Vienna, New York. 233–246

Bronson C.R. (1991): The genetics of phytotoxin production by plant pathogenic fungi. Experientia 47: 771–776

Crute I.R. (1985): The genetic bases of relationships between microbial parasites and their hosts. In: Fraser R.S.S. (ed.): Mechanisms of Resistance to Plant Diseases. Martinus Nijhoff/Dr.W.Junk, Dordrecht, Boston, Lancaster. 81–142

Panaccione D.G. (1993): The fungal genus *Cochliobolus* and toxin-mediated plant disease. Trends Microbiol. 1: 14–20

Pryor T., Ellis J. (1993): The genetic complexity of fungal resistance genes in plants. Advances in Plant Pathology 10: 281–305

Turgeon B.G., Yoder O.C. (1994): Molecular genetics of polyketide toxin biosynthesis by fungi. In: Kohomoto K., Yoder O.C. (eds.): Host Specific Toxin: Biosynthesis, Receptor and Molecular Biology. Faculty of Agriculture, Tottori University, Sogo Printing and Publishing Co., Ltd., Tottori, Japan. 197–205

Wolpert T.J., Navarre D.A., Lorang J.M., Moore D.L. (1996): Evaluation of the glycine decarboxylase complex as the possible site of action of victorin. In: Mills D., Kunoh H., Keen N.T., Mayama S. (eds.): Molecular Aspects of Pathogenicity and Resistance: Requirement for Signal Transduction. The American Phytopathological Society, St. Paul, Minnesota. 244–256

Walton J.D. (1996): Host-Selective Toxins: Agents of Compatibility. Plant Cell 8: 1723–1733

Yoder O.C. (1980): Toxins in pathogenesis. Annu.Rev.Phytopathol. 18: 103–129

Relevant papers

Dewey R.E., Levings III C.S., Timothy D.H. (1986): Novel recombinations in the maize mitochondrial genome produce a unique transcriptional unit in the texas male-sterile cytoplasm. Cell **44**: 439–449

Dewey R.E., Siedow J.N., Timothy D.H., Levings III C.S. (1988): A 13 kilodalton maize mitochondrial protein in *E.coli* confers sensitivity to *Bipolaris maydis* toxin. Science **239**: 293–295

Johal G.S., Briggs S.P. (1992): Reductase activity encoded by the *HM1* disease resistance gene in Maize. Science **258**: 985–987

Navarre D.A., Wolpert T.J. (1995): Inhibition of glycine decarboxylase multienzyme complex by the host-selective toxin victorin. Plant Cell **7**: 463–471

Wolpert T.J., Macko V. (1989): Specific binding of victorin to a 100-kD protein from oats. Proc.Natl.Acad.Sci.USA **86**: 4092–4096

Wolpert T.J., Navarre D.A., Moore D.L., Macko V. (1994): Identification of the 100-kD victorin binding protein from oats. Plant Cell **6**: 1145–1155

Host Plant Resistance:
Cultivar- or Parasite-Specific Resistance

Within a population of susceptible and genetically uniform plants growing in the field and exhibiting basic compatibility with a given pathogen, there may occasionally be found plants that are able to defend themselves against the attacking homologous pathogen. However, a prerequisite for resistance to infection, or incompatibility with a homologous pathogen, is the presence of mutation(s) to resistance in the host plant. The mutation(s) probably arose spontaneously in an ancestral plant but its phenotype became apparent only when the plant was challenged with the appropriate pathogen. In contrast to the broad spectrum of pathogens that are successfully rejected by non-host plants, i.e., by the presence of basic resistance, host resistance is a highly specialized kind of pathogen defense which is directed to only one particular pathogen, or *forma specialis*, or even only to certain races. Except for this one homologous, but now incompatible pathogen, the plant still exhibits full compatibility with other homologous pathogens.

In order to improve agricultural production, during the last 75 years or so, humans have selected cultivars that are resistant to infection by the most injurious phytopathogenic fungi. This newly acquired resistance is highly selective and has been called cultivar-specific resistance.

The classification of cultivars as resistant or susceptible to pathogen attack was first carried out in agriculture. Most farmers grow crop plants of high genetic uniformity in monocultures which may be exposed to populations of homologous pathogens containing different mutants or pathogen races. By general convention, plants grown in genetically uniform monocultures were designated as being resistant to attack by a particular pathogen if they exhibited no or only weak disease symptoms and there were no losses in crop yield, even under conditions of epidemic infection. By contrast, under the same conditions, a susceptible cultivar would show more or less severe disease symptoms resulting in heavy yield losses. In farming the terms resistant and susceptible are used more loosely than in this book where, from the viewpoint of the biologist, resistance means perfect defense of the pathogen thereby preventing both colonization of the plant and propagation of the pathogen. Susceptibility refers to unhampered colonization of the plant with strong propagation of the pathogen leading to the epidemic spread of disease. However, many plant patholgists use resistance to indicate a range of symptoms that includes the production of some limited sporulating lesions up to the point of no macroscopically visible sign of infection or penetration. This latter phenotype is sometimes referred to as immunity. Accordingly, a cereal cultivar called in farming practice disease-resistant may still be colonized, for example, by a rust which produces few spores and is thus unable to spread effectively. This phenomenon has been called "slow rusting". Applying the more strict nomenclature as used in this book, slow rusting would indicate susceptibility of the plant to the pathogen and thus to infection and disease. The behavior of cultivars as resistant or susceptible to pathogen attack, irrespective of the classification used in biology or in agriculture, was originally assumed to be determined only by genes in the host plant. As will be discussed below, this assumption turned out to be correct only for one particular type of host resistance.

Another response observed with crop plants can result in heavy infection by a phytopathogenic fungus which entails no loss in crop yield. In such cases the plant is able to compensate for the damage caused by the phytopathogenic fungus. Accordingly, one speaks of tolerance of the host plant against the pathogen's attack (see also Chap. 10).

Among host plants defined in plant pathology as resistant against pathogen attack, or plants exhibiting at least far-reaching defense against the pathogen, one can discriminate two classes of host resistance which are basically very different. These are race-specific, or vertical resistance and race-non-specific, or horizontal resistance. Both will be discussed in detail in the next sections (race-specific in Sects. 9.1–9.1.8, race-non-specific in Sect. 9.2). Here the main characteristics of both types will be characterized only briefly.

Race-specific or vertical resistance is monogenically inherited, and controlled by "major genes". This term, often used in plant breeding, refers to genes causing strong phenotypic effects and which segregate in the progeny of crosses between resistant and susceptible plants as two easily discernible phenotypes (see Sect. 9.1.1). Race-specific resistance has also been described as qualitative resistance since it is a qualitative character that is either fully present or absent.

In contrast, race-non specific or horizontal resistance is polygenically determined, by the combined action of many "minor genes" of individually small effect. The additive action of these genes in phenotypic expression is a quantitative character described as quantitative resistance. Among the progeny of crosses between such a resistant plant and a susceptible cultivar one finds many different phenotypes with respect to strength of resistance, but the strength of each one of the segregants is generally less than that of the resistant parent (see Sect. 9.2). The discrimination between "major" and "minor" genes characterizes only their different phenotypic strengths, i.e., expression of either a strong or a weak effect.

Cultivar specific or vertical resistance, determined by major genes, is expressed qualitatively and "discontinuously", implying that this phenotype represents an "either/or" decision. Either there ensues complete defense by the plant against the pathogen (incompatible reaction), or no defense at all leading to unhampered colonization by the parasite (compatible reaction). In contrast, minor genes, that determine horizontal resistance, are expressed quantitatively and additively, the degree of resistance against pathogen attack increases "continuously" with increasing numbers of minor genes engaged.

However, there is another, very important difference between race-specific and race-non-specific resistance, that concerns their genetic determination. Race-non-specific resistance is determined by the host plant alone, while race-specific resistance is determined by both the host plant and the pathogen (for details see Sect. 9.1.1). This may have serious consequences for farming since the race-specific resistance of particular cultivars may disappear after several seasons, because specific virulent pathogen mutants (see Sect. 9.1.1) may be present in the infecting pathogen population. Since race-specific resistance depends on genetic determinants in the host plant and the pathogen, a single mutation to specific virulence in the pathogen population may result in the disappearance of the expression of race-specific resistance among the host plants. Within a few reproductive cycles selective propagation of the specific virulent pathogen mutant will displace all the avirulent pathogens in the population which are repelled by the resistant host. Thus, race-specific resistance may appear as a rather unstable genetic trait, because the pathogen can overcome it. By contrast, race-non-specific resistant cultivars do not lose their host resistance, because it is determined solely by the host plant. In other words, pathogens are unable to overcome race-non-specific resistance by a single mutational step. Rather, this would require accumulation of many mutational steps in the pathogen to develop new pathogenicity genes. This may well require very long periods of time and be feasible only in the course of evolution. Race-non-specific resistance is a trait that is genetically very stable.

Race-specific or vertical resistance represents the plant's secondary means of defending itself against an attacking pathogen. For this purpose, the plant employs the same repertoire of defense mechanisms as when expressing active basic resistance. Race-specific resistance is superimposed on basic compatibility, i.e. on reaction chains effecting colonization of the plant, however, without turning them off. This was demonstrated by Michèle Heath (1974, 1981) for the rust fungus *Uromyces phaseoli* var. *vaginae* interacting either with non-host plants or with its host *Phaseolus vulgaris*. Employing electron microscopic techniques she showed that the defense reactions of race-specific resistance are expressed *after* the defense reactions of basic resistance. Since the appearance of race-specific resistance in correspondingly resistant plants requires basic compatibility, i.e., after basic resistance has been overcome, establishment of race-specific resistance is subordinate to basic resistance.

Race-specific resistance has been investigated very thoroughly because of its relatively simple experimental accessibility, its easily monitored genetic determination, and, in many cases, also its clear cut "yes or no" expression of resistance. According to recent experiments, processes involved in triggering race-specific resistance, particularly the signal-sensor reactions for recognition and the signal transduction events that follow, may also operate in basic resistance.

It should also be recalled that recognition between a host-selective toxin and its receptor in the susceptible host plant (see Chap. 8, Sect. 8.2) proceeds as a signal-sensor reaction. Evidently signal-sensor reactions play an important role in communication between plants and their pathogens as well as well as within and between the plant's own cells. Thus, experiments which originally aimed at investigating the molecular nature of the trigger for expression of race-specific resistance now extend far beyond the somewhat narrow frame of plant pathology into basic research in botany and general biology.

Comparisons of host resistance among different cultivars reveal considerable variation in their efficiency in preventing plant disease. Thus examples of pathogen-specific defense, even when expressed, as in race-specific resistance, as a "yes or no" alternative, are often neither complete nor alike among different cultivars. Disease symptoms may appear, even after an incompatible interaction, indicating that resistance against the pathogen was incomplete. The term "partial resistance" has been coined for such more or less incomplete pathogen-specific defense. Partial resistance may range from more or less complete resistance to nearly full susceptibility. Complete, or almost complete, pathogen defense is found frequently in race-specific or vertical resistance against biotrophic pathogens. Race-specific resistance that is partial is much less common (see also Sect. 9.1). In contrast, one finds as the rule in race-non-specific or horizontal resistance only partial resistance, its degree depending on the number of genes acting additively. The disease resistance of non-host plants, their basic resistance being effective against non-homologous pathogens, is always complete, never partial.

Reviews

Bailey J.A. (1983): Biological perspectives of host-pathogen interactions. In: Bailey J.A., Deverall B.J. (eds.): The Dynamics of Host Defence. Academic Press Australia, Sidney, New York, London, Paris, San Diego, San Francisco, Sao Paulo, Tokyo, Toronto. 1–32

Dixon R.A., Harrison M.J. (1990): Activation, structure, and organization of genes involved in microbial defense in plants. Adv.Genetics 28: 165–234

Dixon R.A., Lamb C.J. (1990): Molecular communication in interactions between plants and microbial pathogens. Annu.Rev.Plant Physiol. 41: 339–367

Ellingboe A.H. (1976): Genetics of host-parasite interactions. In: Heitefuss R., Williams P.H. (eds.): Encyclopedia of Plant Physiology, (NS), Physiological Plant Pathology. Springer Verlag, Heidelberg. 761–778

Ellingboe A.H. (1981): Changing concepts in host-pathogen genetics. Annu.Rev.Phytopathol. 19: 125–143

Gabriel D.W., Rolfe B.G. (1990): Working models of specific recognition in plant-microbe interactions. Annu.Rev.Phytopathol. 28: 365–391

Heath M.C. (1991): Evolution of resistance to fungal parasitism in natural ecosystems. New Phytol. 119: 331–343

Johnson R. (1992): Past, present and future opportunities in breeding for disease resistance, with examples from wheat. Euphytica 63: 3–22

Keen N.T. (1990): Gene-for-gene complementarity in plant-pathogen interactions. Annual Review of Genetics 24: 447–463

Keen N.T. (1992): The molecular biology of disease resistance. Plant Mol.Biol. 19: 109–122

Keen N.T. (1993): An overview of active disease defense in plants. In: Fritig B., Legrand M. (eds.): Mechanisms of Plant Defense Responses. Kluwer Academic Publishers, The Netherlands. 3–11

Parlevliet J.E., Zadoks J.C. (1977): The integrated concept of disease resistance: A new view including horizontal and vertical resistance in plants. Euphytica 26: 5–21

Vanderplank J.E. (1963): Plant Diseases: Epidemics and Control. Academic Press, New York, London

Vanderplank J.E. (1968): Disease Resistance in Plants.. Academic Press, New York, London

Relevant papers

Heath M.C. (1974): Light and electron microscope studies of the interactions of host and non-host plants with cowpea rust – *Uromyces phaseoli* var.*vignae*. Physiol.Plant Pathol. 4: 403–414

Heath M.C. (1981): Resistance of plants to rust infection. Phytopathology 71: 971–974

9.1
Race-Specific, or Vertical, Resistance

Flor's pioneering work on the genetics of resistance in cultivated plants against attack by phytopathogenic fungi, started in the early 1930s. His first publication in this field appeared in 1935, and in subsequent years many papers followed analyzing what we now call the race-specific or vertical resistance of plants that are hosts for certain pathogenic fungi. To study genetic determination of infection susceptibility and resistance, Flor performed his classic crossing experiments among susceptible and resistant flax cultivars and among virulent and "non-virulent" races of flax rust. The results of his experiments led Flor to postulate the so called gene-for-gene hypothesis for determination of race-specific resistance. The exciting point of his experiments was the finding that this kind of resistance is genetically determined in both partners. In other words, both host plant and pathogen determine together, each with a particular gene, the phenotype of plant resistance. This suggested that something like recognition takes place between the host plant and its pathogen. In the following sections, the phenomenon of race-specific resistance will be discussed in general terms, the next will deal in more detail with the genes involved.

Recognition of a particular pathogen race by a corresponding resistant cultivar has been attributed to the existence of two corresponding factors recognizing each other. That synthesized by the host plant, was called a resistance factor, and the other, derived from the pathogen, an avirulence factor. The hypothetical existence of both factors and their mutual recognition formed the basis for further investigations. The recognition must be highly specific, since only factors that correspond to each other trigger the expression of race-specific resistance. Recognition may also be described in the following way: A single resistance factor in the host plant effects expression of defense reactions only if the infecting pathogen produces a corresponding avirulence factor which this particular resistance factor is able to recognize. This implies that the correspondence between the plant's resistance factor and the pathogen's avirulence factor triggers expression of defense reactions and prevents the pathogen from parasitizing its host plant. Under such conditions the pathogen is called avirulent. In contrast, the pathogen is virulent if it is *not* recognized by its host plant. The plant is parasitized because it lacks the corresponding resistance factor.

The interaction between a phytopathogenic fungus and its host plant may be briefly described as follows: If the pathogen has overcome the basic resistance of the plant, thus establishing basic compatibility, parasitic colonization by the pathogen ensues. This colonization may be prevented secondarily if the plant harbors a mutation to race-specific resistance which effects the production of a particular resistance factor able to recognize a particular corresponding avirulence factor synthesized by the pathogen. The ensuing recognition between the corresponding factors triggers expression of defense reactions, i.e., synthesis by the plant of new defense barriers.

The secondarily established race-specific resistance of plants is observed mostly with biotrophic pathogens, because they keep the infected host plants alive and metabolically active in order to feed on them. However, if an infected plant carries a mutation to race-specific resistance, enabling it to recognize pathogens of one particular race, it can build new defense barriers employing its own metabolism.

The defense reactions expressed by different race-specific resistant cultivars against obligate biotrophs such as rusts and mildews may vary in strength depending on the particular resistance mutation involved. Also, physiological conditions in the host plant may modify expression of resistance. Not only may different degrees of resistance be observed, i.e., partial resistance, as discussed above (Chap. 9), but also different resistance phenotypes. Nevertheless, in all of these cases the preceding recognition reaction triggering expression of the different resistance responses is of an "all or none" type. Thus, the spectrum of race-specific defense reactions may vary quantitatively: At one extreme there is complete incompatibility resulting in the appearance of hypersensitive necrotic spots preventing growth and reproduction of the pathogen that are hardly visible macroscopically. This is the most common response in the race-specific resistance of host plants. At the other extreme we may observe only "weak" incompatibility, with limited pathogen growth in the plant permitting some spore production, allowing some spread of the pathogen within the plant population.

The degree of incompatibility is influenced by many different conditions, for example, how fast or complete the host plant's defense reactions limit pathogen nutrition and its success in colonizing the plant. The pathogen may be obstructed by formation of callose plugs at penetration sites, that impede development of haustoria for pathogen nutrition. Thus a "race" between competing reaction chains in the plant and the pathogen begins that leads either to compatibility or incompatibility, i.e., either more or less effective growth and reproduction of the pathogen or a complete blockade of it's growth and spread. The progress and result of this competition leads to differences in morphology and strength of defense reactions. The degree of race-specific resistance has been experimentally measured by comparing the frequencies of variably developed necroses and other symptoms observed on resistant and susceptible host plants. Thus on plants diseased with rusts or mildews five different infection types can be seen (see Sect. 9.1.1).

The result of the competing reaction chains activated after infection of race-specific resistant plants with an avirulent pathogen may, in rare cases, remain undecided for a long period of time. This has been observed, with *Puccinia graminis* infecting certain race-specific resistant and race-non-specific resistant cultivars (see Sect. 9.2). The infecting pathogen can, despite its strongly restrained vegetative growth, still produce a few uredospores, allowing for slow spreading of the disease. This has been called slow rusting. Cereal cultivars with slow rusting symptoms are considered in agricultural practice as "partially resistant". This means the diseased cultivar suffers only negligible losses in yield but sustains latent spreading of the pathogen. The phenotypes of race-specific resistance can be modified by experimental manipulation. For example when a race-specific resistant wheat cultivar inoculated with *Puccinia graminis* f.sp. *tritici* is treated with ethylene, the expected incompatible interaction is converted into a compatible one (Daly et al. 1970). This is an extreme example of the competition between the two reaction chains where experimental manipulation shifts it to favour parasitization.

When plants are attacked by necrotrophic pathogens, which as a rule immediately kill the infected plant, the situation is quite different. Even if the host plant carries a mutation to race-specific resistance there is no chance for the plant to build up new defense reactions, because its metabolism would be demolished. However, some rare exceptions have been found from the rule of immediate killing with the rye pathogen *Rhynchosporium secalis*. Plant cells carrying a mutation to race-specific resistance against this pathogen are still able to build up defense barriers, because the pathogen does not kill the host cells immediately after infection. In this way there is a chance for the infected cell, in spite of existing basic compatibility, to build up secondary defense barriers (see also Sect. 9.1.3.1).

As mentioned already, the battery of defense reactions engaged in race-specific resistance resemble those in active basic resistance (see Chap. 5). Therefore, the only difference between active basic resistance and race-specific resistance seems to be in the recognition systems that trigger their expression and possibly also in some of the ensuing signal transduction processes, but certainly not in the type of defense reactions expressed. Thus, there are two groups of genes with quite different functions that are engaged on two different levels in plant defense reactions. They are the genes that determine the recognition between plant and pathogen and the genes that control the expression of plant defense reactions. These latter reactions are common to both active basic resistance and race-specific resistance. In other words, the plant has access to the same defense reactions via two different recognition events, those triggering active basic resistance and those triggering race-specific resistance. In describing race-specific resistance, the plant genes engaged in recognizing attacking pathogens are called "resistance genes" and the genes coding for the different defense reactions are called "disease resistance response genes" or "defense response genes".

Host plants expressing race-specific resistance produce the resistance factor only if they carry the gene coding for it. They can aquire it by mutation or inherit it from a resistant parent. Cultivars of crop plants are by definition not wild types. For the most part the race specific resistance genes they carry were introduced by plant breeders. In contrast, wild type pathogens, that have not encountered crop cultivars, may produce many different avirulence factors, implying that avirulence factors represent wild phenotypes. Furthermore, a plant may carry many different genes for race-specific resistance which map at several different chromosomal loci, but their total number is much smaller than the number of genes coding for potential avirulence factors in the pathogen (see Sect. 9.1.5).

If a host plant has several different genes coding for race-specific resistance and thereby effects a hypersensitive reaction, the interaction of that plant with a pathogen race may lead to the formation of more than one pair of resistance/avirulence factors. This would allow several different, specific recognition events triggering hypersensitive reactions to occur at the same time. However, different specific recognition events occurring in the same plant do not show an additive effect boosting the degree of race-specific resistance against pathogen attack. This is because the qualitative "either/or" character, of a single recognition event is sufficient for optimal resistance expression. Nonetheless, the presence of several race-specific resistance genes in one cultivar will increase the plant's ability to defend itself against several different pathogen races.

Finally, one may summarize as follows: Two types of genes coding for quite different functions cooperate in pathogen defense controlled by race-specific resistance: (1) resistance genes, the products of which recognize the pathogen to be repelled, and (2) defense response or disease resistance response genes that express defense mechanisms. Recognition of a homologous pathogen triggers, via signal transduction events, the expression of defense response genes. Hence, the expression of defense response genes is regulated, via signal transduction, by resistance genes.

Reviews

Bailey J.A. (1983): Biological perspectives of host-pathogen interactions. In: Bailey J.A., Deverall B.J. (eds.): The Dynamics of Host Defence. Academic Press Australia, Sidney, New York, London, Paris, San Diego, San Francisco, Sao Paulo, Tokyo, Toronto. 1 – 32
Burnett J.H. (1975): Chapter 13: General aspects of fungal pathogenicity. In: Burnett J.H. (ed.): Mycogenetics. John Wiley & Sons, London, New York, Sidney, Toronto. 259 – 287
Crute I.R. (1985): The genetic bases of relationships between microbial parasites and their hosts. In: Fraser R.S.S. (ed.): Mechanisms of Resistance to Plant Diseases. Martinus Nijhoff/Dr.W.Junk, Dordrecht, Boston, Lancaster. 81 – 142
Dixon R.A., Lamb C.J. (1990): Molecular communication in interactions between plants and microbial pathogens. Annu.Rev.Plant Physiol. 41: 339 – 367
Ellingboe A.H. (1976): Genetics of host-parasite interactions. In: Heitefuss R., Williams P.H. (eds.): Encyclopedia of Plant Physiology, (NS), Physiological Plant Pathology. Springer Verlag, Heidelberg. 761 – 778
Flor H.H. (1971): Current status of the gene-for-gene concept. Annu.Rev.Phytopathol. 9: 275 – 296
Gabriel D.W., Rolfe B.G. (1990): Working models of specific recognition in plant-microbe interactions. Annu.Rev.Phytopathol. 28: 365 – 391
Heath M.C. (1991): Evolution of resistance to fungal parasitism in natural ecosystems. New Phytol. 119: 331 – 343
Keen N.T. (1990): Gene-for-gene complementarity in plant-pathogen interactions. Annual Review of Genetics 24: 447 – 463
Keen N.T. (1992): The molecular biology of disease resistance. Plant Mol.Biol. 19: 109 – 122
Keen N.T. (1993): An overview of active disease defense in plants. In: Fritig B., Legrand M. (eds.): Mechanisms of Plant Defense Responses. Kluwer Academic Publishers, The Netherlands. 3 – 11
Person C. (1959): Gene-for-gene relationships in host : parasite systems. Can.J.Bot. 37: 1101 – 1130
Pryor T. (1987): The origin and structure of fungal disease resistance genes in plants. Trends Genet. 3: 157 – 161

Relevant papers

Daly J.M., Seevers P.M., Ludden P. (1970): Studies on wheat stem rust resistance controlled at the Sr6 locus. III. Ethylene and disease reaction. Phytopathology 60: 1648 – 1652

9.1.1
The "Gene-for-Gene" Hypothesis

To comprehend the basis of race-specific resistance it is helpful to understand some aspects of its genetic determination such as the number of genes involved. As already discussed in the last chapter, in populations of susceptible host plants, mutants able to resist the attack of a particular pathogen species or race may occasionally appear. The discovery of such race-specific resistant mutants appeared to be a promising start for breeding infection-resistant cultivars for agricultural use. However, after they were grown in the field for several seasons, the plants lost the ability to defend themselves against the pathogens they were initially resistant against. It soon became apparent that in the attacking pathogen population new mutants may appear relatively frequently, that negate the resistance of their host plants and thereby become able to colonize them. These newly emerged pathogen mutants were designated as being "virulent", in contrast to the "avirulent" pathogen types which the race-specific resistant cultivars were effective against. These virulent mutants were classified as belonging to different "physiological races" of the pathogen. Thus, in both the host plant as well as the homologous pathogen, genetic determinants may decide if their interaction is incompatible or compatible, indicating whether the race-specific resistant plant is able to defend itself against, or whether it becomes parasitized by a particular pathogen race.

Between 1940 and 1960 Flor carried out crossing experiments between different physiological races of the rust *Melampsora lini*, a biotrophic pathogen on flax, and different cultivars of its host plant *Linum usitatissimum* either susceptible or resistant to rust infection (Flor 1942, 1946, 1955). Similarly designed experiments were subsequently performed by other research groups, who also used other cultivated plants such as cereals parasitized by *Puccinia* spp.. However, very few were able to study the genetic controls in both the host and the parasite.

Before explaining the crossing experiments performed by Flor, some details about the biology and genetics of flax rust will be briefly discussed. *Melampsora lini*, a basidiomycete exhibits during its life cycle two different developmental phases. In the first, the pathogen's mycelium consists of haploid cells which form a monokaryon. In the second phase, each cell contains two different haploid nuclei which form a dikaryon. The dikaryon is the result of sexual mating between two monokaryotic mycelia of different mating type at the end of the first developmental phase. The change between these two developmental phases is referred to as the alternation of generations. Both, the monokaryotic and dikaryotic mycelia parasitize the same host plant, *Linum usitatissimum*. (This autoecious life style observed with flax rust contrasts to the much more frequently observed heteroecious life style exhibited, for example, by the cereal rust *Puccinia graminis*, whose haploid mycelium parasitizes *Berberis vulgaris*, *barberry*, or

---

→

Figure 3. Flax rust, *Melampsora lini*, a fungal parasite of *Linum usitatissimum*, life cycles and crossings: The life cycle, consisting of two developmental phases, a haploid and a dikaryotic one, is schematically described in the *left* part of the figure. In the *right* part of the same figure two crosses are schematically depicted between genetically marked dikaryotic parental pathogens (P) yielding either F_1- or F_2 progeny. The haploid gametes formed by the dikaryotic mycelium correspond to genes (or alleles), the segregation of which shall be traced. The gametes can be combined freely for distribution into the progeny generation, determining the genotype of the dikaryotic mycelium. Hence, the latter contains genes from both parents. Each of the haploid heterothallic mycelia is grown separately on a susceptible plant. Testing of the parental pathogens and their F_1 and F_2 progenies for avirulence or virulence is performed on corresponding race-specific resistant plants (differentials). – A situation is depicted in which genotypes and phenotypes obtained in the F_1 and F_2 progenies correspond to genetic determination of avirulence by a single dominantly expressed gene, Avr_1, interacting with a R_1 race-specific resistant host plant

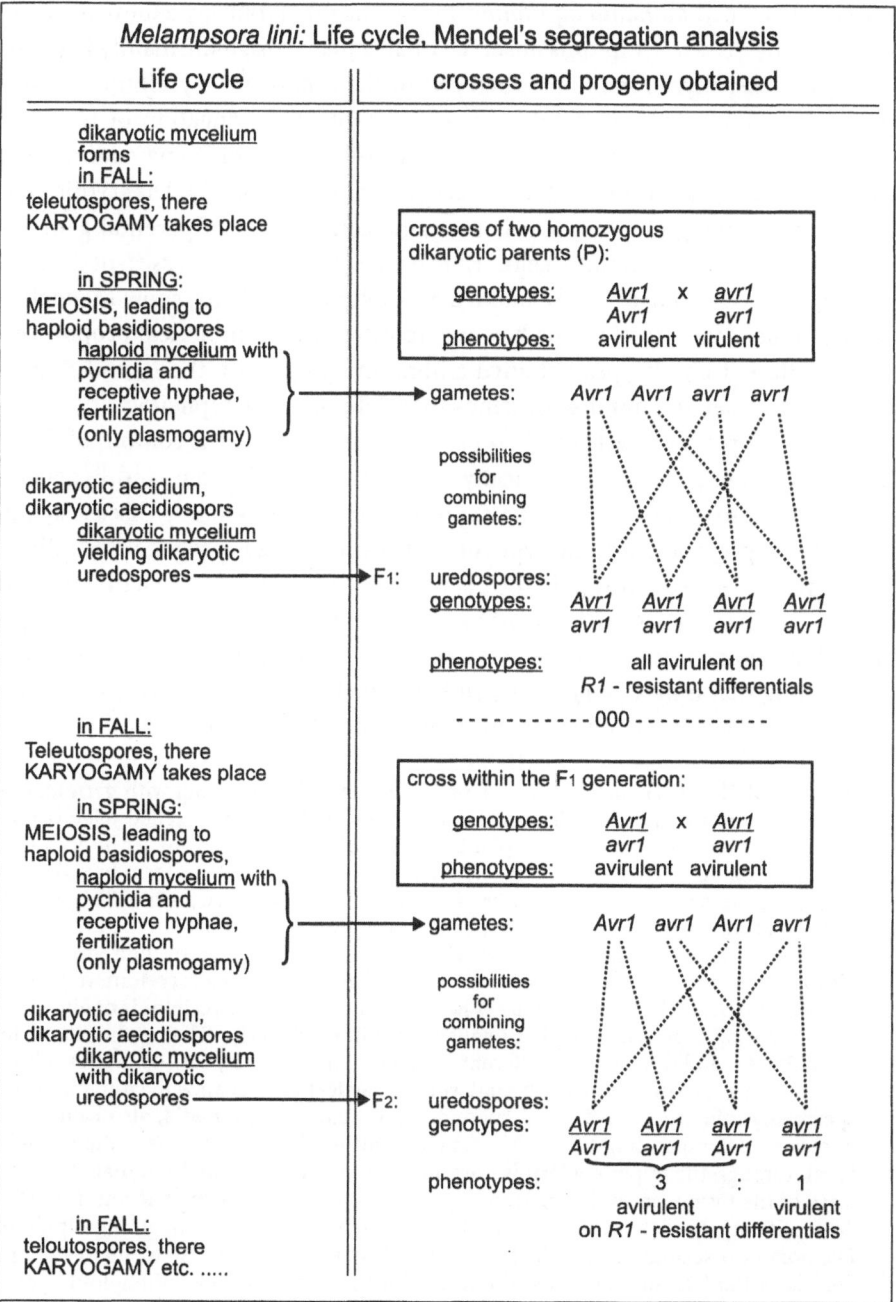

Melampsora lini: Life cycle, Mendel's segregation analysis

Life cycle	crosses and progeny obtained

Life cycle:

dikaryotic mycelium
forms
in FALL:
teleutospores, there
KARYOGAMY takes place

in SPRING:
MEIOSIS, leading to
haploid basidiospores
haploid mycelium with
pycnidia and
receptive hyphae,
fertilization
(only plasmogamy)

dikaryotic aecidium,
dikaryotic aecidiospors
dikaryotic mycelium
yielding dikaryotic
uredospores

in FALL:
Teleutospores, there
KARYOGAMY takes place

in SPRING:
MEIOSIS, leading to
haploid basidiospores,
haploid mycelium with
pycnidia and
receptive hyphae,
fertilization
(only plasmogamy)

dikaryotic aecidium,
dikaryotic aecidiospores
dikaryotic mycelium
with dikaryotic
uredospores

in FALL:
teleutospores, there
KARYOGAMY etc. .....

crosses and progeny obtained:

crosses of two homozygous
dikaryotic parents (P):

genotypes: *Avr1* x *avr1*
              Avr1   *avr1*
phenotypes: avirulent virulent

gametes: *Avr1 Avr1 avr1 avr1*

possibilities
for
combining
gametes:

F₁: uredospores:
genotypes: *Avr1*   *Avr1*   *Avr1*   *Avr1*
             avr1   *avr1*   *avr1*   *avr1*

phenotypes: all avirulent on
              R1 - resistant differentials

- - - - - - - - - - 000 - - - - - - - - - - -

cross within the F₁ generation:

genotypes: *Avr1* x *Avr1*
              avr1   *avr1*
phenotypes: avirulent   avirulent

gametes: *Avr1 avr1 Avr1 avr1*

possibilities
for
combining
gametes:

F₂: uredospores:
genotypes: *Avr1*   *Avr1*   *avr1*   *avr1*
             Avr1   *avr1*   *Avr1*   *avr1*

phenotypes: 3   :   1
              avirulent   virulent
              on *R1* - resistant differentials

the closely related *Mahonia aquifolia* whereas the dikaryotic mycelium grows on cereals.) The disease symptoms observed on flax plants originate mainly from the dikaryotic rust mycelium. Since each cell of the dikaryotic mycelium contains nuclei from both parents, the dikaryon represents a prolongation of the stage between fusion of the two parental cells (plasmogamy or somatogamy) and fusion of their haploid nuclei (karyogamy) leading to diploidy. Dikaryotic cells contain the same gene pool, that is the same complement of parental genes as would be present in diploid cells. Hence, the progeny of crosses performed among dikaryotic fungal mycelia, like their diploid host plants, show Mendelian segregations. After crossing one has only to estimate the numerical proportions of the different phenotypes obtained among the progeny in the F_1 and F_2 generations, to infer the number of genes that determine each phenotype.

The developmental cycle of *M. lini* is summarized in the left column of *Fig. 3*. The right column shows schematically crosses that give rise to F_1 and F_2 generations leading to dikaryotic mycelia expressing the corresponding disease symptoms. The segregation of phenotypes observed among the F_2 progeny can indicate the number of genes involved in the phenotype of race-specific resistance and the mode of gene expression, dominant or recessive. *Figure 3* demonstrates segregation of the characters avirulence and virulence – more strictly, specific virulence – assuming that only a single gene, with alleles Avr_1 (dominant) or avr_1 (recessive), is involved in the determination of avirulent behavior.

Life cycle of M. lini. The haploid phase of *M. lini* starts in the spring with germination of the haploid **basidiospore**. The rust overwinters as diploid, two-celled **teleutospore** (teliospore) on debris of the host plant. In the spring the diploid nuclei in each teleutospore cell undergo meiosis. The teleutospores then germinate to form **basidia** each of which produces four basidiospores or haploid meiotic products, two of "+" and two of "–" **mating type**. The basidiospores are discharged and germinate in the presence of water. If the spore has landed on the surface of a flax leaf the germ tube of the basidiospore penetrates the upper epidermis to form within the leaf a mycelium of haploid uninucleate cells. The **haploid phase** is **heterothallic** and each mycelium is of either "+" or "–" mating type. The mycelial hyphae grow intercellularly, forming haustoria for nourishment and develop two different sexual organs necessary for the pathogen's reproduction: (1) close to the upper epidermis of the leaf flask-shaped **pycnidia** in which **spermatogonia** are formed. Each spermogonium cuts off **spermatia**, also called **pycnospores**, corresponding to "male" gametes. During their development the pycnidia break through the upper epidermis, forming an opening, the **olpidium**, that allows release of the pycnospores. Furthermore, pycnidia develop sterile hyphae and branched "**receptive hyphae**", with no cross walls, that extend outwards through the olpidium. Pycnidia also secrete nectar which is collected by insects. (2) Close to the lower epidermis of the leaf, the same mycelium develops round aggregates of haploid hyphae forming precursors of **aecidia**. They harbor so-called haploid **basal cells**, the nuclei of which correspond to "female" gametes. The first step in the process of sexuality is fertilization occuring outside the pycnidium and consists of fusion between receptive hyphae and pycnospores. However, this somatogamy is *not* soon followed by a karyogamy. Instead the pycnospore's nucleus migrates down the receptive hypha, traversing the cross-walls of further mycelial cells, finally reaching the basal cell within the precursor aecidium. In this way a **dikaryotic phase**, i.e. a **dikaryon** is established, forming first an **aecidium** which then starts forming spores. During its development the aecidium breaks through the lower epidermis of the leaf, releasing the dikaryotic **aecidiospores** (aeciospore) that have formed inside the aecidium in long chains. The aecidiospores are spread by wind, infecting more distant leaves and initiating formation of new dikaryotic mycelia. The secondary dikaryotic mycelia produce **uredinia** which form further generations of dikaryotic spores, the **uredospores**, which facilitate very extensive pathogen propagation by wind.

Self-fertilization (i.e., somatogamy) within a pycnidium is excluded since mating occurs only between pycnospores and receptive hyphae of different mating types. Under natural conditions transport of pycnospores between pycnidia is accomplished by insects collecting the pycnidium's secreted nectar and carrying the pycnospores with them.

Only in climates with cold winters do thick-walled **teleutospores** develop in **telia** on leaf surfaces of flax during the fall. The two-celled teleutospores remain at first dikaryotic, but during hibernation each cell undergoes karyogamy, leading to two-celled, diploid teleutospores. In the spring, after germination of the teleutospores and production of basidiospores a new developmental cycle starts (see above). In some other rust fungi vegetative propagation may continue in an unlimited manner only within the dikaryotic phase, i.e., without changing to a haploid phase allowing for expression of sexuality providing for occurrence of genetic recombination.

It should be stressed that, during the haplophase of flax rust, no clonal, or vegetative propagation occurs, since the pycnospores serve only a sexual function, namely allowing for increasing genetic variability: As "male" gametes, the pycnospores function only in fertilizing "female" gametes, the receptive hyphae, leading via the diakaryotic phase finally to karyogamy and meiosis allowing for genetic recombination. In contrast, true clonal proliferation without genetic recombination takes place during the dikaryotic phase by the production of aecidio- and uredospores. Accordingly, it has been shown that in pathogens having a sexual stage like rusts (*Puccinia spp.*), smuts (*Ustilago* spp., *Tilletia* spp.), and apple scab (*Venturia inaequalis*) hybridization is an important means of pathogenic variation. Furthermore, the haplophase of flax rust expresses the same genes that govern pathogenicity and virulence in the dikaryon.

Crosses between pathogen races: To perform crosses between different pathogen races (Flor 1942) telia were collected with teleutospores of the corresponding parental types. The germinating teleutospores were induced to develop basidiospores and these were used for infecting susceptible host plants. Each basidiospore germinating on a flax leaf gives rise to a haploid mycelium. The experimental manipulation for performing a particular cross is the following: Employing a sterile wire loop, pycnospores derived from plant A infected with basidiospores of mating type "+" are transferred to receptive hyphae of a pycnidium on plant B infected with a "−" mating type. However, such crosses will be fertile only when pycnospores and receptive hyphae are of opposite mating type. The aecidiospores or uredospores obtained in the following dikaryotic generation (see scheme in *Fig. 3* , right column, "F_1") represent genotypes exactly corresponding to the F_1 progeny obtained from crossing diploid plants. To obtain F_2 progeny, susceptible plants were infected with the F_1 aecidiospores (or uredospores) and teleutospores were collected in the fall. During the following spring, the basidiospores developed from the teleutospores give rise to haploid mycelia which then are crossed as described above, yielding F_2 aecidiospores (or uredospores) (see *Fig. 3*, right column, "F_2"). The dikaryon genotypes obtained result from random combinations among gametes formed by the parents (*Fig. 3*, right column).

Identification of avirulent and virulent phenotypes among the pathogen progeny: The dikaryotic aecidio- and uredospores of the F_1 and F_2 generations were tested for virulence or avirulence on particular tester plants, the so-called **differentials**. These represent a collection of resistant flax plants, each one carrying a different gene for race-specific resistance. Differentials consist of pure lines of homozygous cultivars multiplied by selfing (flax is self-fertile). Each line carries only a single determinant for race-specific resistance which segregates in crosses like a single locus, suggesting that only one gene determines this trait. However, one reservation should be kept in mind: each race specific resistance gene can only be detected if the "corresponding" avirulence gene is present in the pathogen (see below, paragraph (1)). As an example, Flor showed that Bison, his universal susceptible flax cultivar in fact carried gene(s) for resistance to races of *M. lini* not normally found in N.America.

For genetic characterization of a dikaryotic aecidiospore, a set of differentials is infected with uredospores derived from this single aeciidiospore. Then the type of infection that develops on each differential, incompatible or compatible, is recorded. Incompatible or compatible responses indicate whether the pathogen carried the "corresponding" *Avr* gene or its allele *avr*. Determining the race specific genes for resistance of cultivars against known pathogen races follows the same principle: The genotypes obtained in the F_1 and F_2 progeny from crosses between different cultivars are identified by inoculating them with a set of characterized pathogen races. The outcome, compatible or incompatible, allows identification of the resistance specificity carried by a flax cultivar.

The rust/flax system offers some particular advantages for analyzing the genetic determination of race-specific resistance. The features of this system, some of which were already mentioned, may be summarized as follows:

(a) After crossing, both partners of the system, the dikaryotic pathogen *M. lini* and its diploid host plant *L. usitatissimum*, can be analyzed for mendelian segregation: The frequencies of different phenotypes obtained in the F_1 and F_2 progenies indicate the numbers of genes determining avirulence and virulence in the pathogen and resistance and susceptibility in the host plant.

(b) Mycelia of both the haploid and the dikaryotic developmental phases grow on the same plant, flax (autoecious lifestyle).

(c) Successful infection and colonization of *L. usitatissimum* by haploid and dikaryotic mycelium of *M. lini* and host plant resistance are easy to score. Nevertheless the expression of resistance reactions and the extent of plant colonization by the parasite may vary among different cultivars of flax exhibiting race-specific resistance. The variation depends mainly on the competition between the two opposing reaction chains discussed above (Chap. 9, Sect. 9.1), that are directed either towards resistance or plant colonization. Therefore, when scoring for infection responses, Flor had to differentiate between five morphologically different infection types: a hypersensitive reaction (Flor's reaction class 0), small- to medium-sized uredinia surrounded by a necrotic area (reaction classes 1 and 2), and high to very high infection susceptibility displaying medium- to large-sized uredinia surrounded by modest or more strongly pronounced chloroses but without any necroses (reaction classes 3 and 4). Classes 0, 1, and 2 were considered resistant reactions, and classes 3 and 4 as susceptible reactions.

(d) *L. usitatissimum* is self-fertilizing so that it is easy to select and maintain pure homozygous lines of differentials for identifying virulence and avirulence markers carried by pathogens.

(e) Flax plants are particularly suitable for experiments that require the testing of each plant with a number of different pathogen cultures, at the same time and under the same physiological conditions. Successive leaves of the same stem can be employed for this purpose recording infection responses after inoculation with aecidio- or uredospores.

The evaluation of his crosses led Flor to the following conclusions:

(1) The appearance of an incompatible interaction, i.e., effective resistance against the attacking pathogen, depends on the presence of one particular resistance gene R in the host plant and on another particular avirulence gene Avr in the pathogen and both these genes must "fit" each other. In other words, some kind of "correspondence" between both genes is required for releasing the expression of race-specific resistance in the plant. Take for example the corresponding gene pair $R_1 - Avr_1$ (corresponding gene pairs in host plant and pathogen are separated by a hyphen and marked by the same indices. e.g., $R_x - Avr_x$, whereas alleles of the same gene in a diploid or dikaryotic organism are marked by a slash, e.g., R_1/R_1, R_1/r_1, or Avr_1/Avr_1, Avr_1/avr_1): Each gene may be found in the diploid plant as homozygous R_1/R_1 (for resistance) or r_1/r_1 (for susceptibility) alleles and in the dikaryotic pathogen as Avr_1/Avr_1 or avr_1/avr_1 alleles. Four different allele combinations may be found when both partners interact, resulting in the responses shown in Table 6. From this so-called quadratic check, it follows that only the allele combination $R_1 - Avr_1$ triggers an incompatible interaction, all other combinations allow parasitism of the plant, i.e., they yield compatible interactions. (In Table 6 it is assumed that the pathogen is a dikaryotic organism, such as *M. lini*. In other cases the pathogen may be haploid, such as the ascomycete *Magnaporthe grisea*, which infects rice plants. However, this does not alter the fact that a single gene pair, for example $R_1 - Avr_1$ suffices to release incompatibility.) A single race-specific resistant plant may contain one or several race-specific resistance genes all of which we may assume came into being by mutation. By contrast, avirulence genes may be regarded as wild type alleles and we may suppose that pathogens harbor many different ones.

(2) Gene R_1 determining race-specific resistance, is dominant over r_1 for susceptibility to infection. The pathogen's avirulence gene, Avr_1, is dominant over its mutant allele avr_1, thereby bringing about avirulence. Therefore, an R_1/r_1 heterozygous plant expresses only allele R_1 and a heterozygous dikaryotic mycelium Avr_1/avr_1 expresses only the wild type allele Avr_1. This may be demonstrated by crossing two dikaryotic pathogens, Avr_1/Avr_1 avr_1/avr_1, yielding an F_1 progeny with uniformly avirulent phenotypes (genotypes Avr_1/avr_1). After crossing among these, the F_2 progeny exhibits a 3 : 1 segregation of avirulent (genotypes Avr_1/Avr_1 and avr_1/Avr_1) to virulent phenotypes (genotype avr_1/avr_1) (see *Fig. 3*,

lower right). The dominance of the R_1 and Avr_1 alleles is discussed below in relation to the quadratic check.

The experimental results stated under (1) constitute the basis for the gene-for-gene hypothesis, which describes the genetic determination of race-specific resistance. Generally, the hypothesis postulates that one single gene for resistance in the host plant, R_1, and another single gene for avirulence in the pathogen, Avr_1, are required for expression of the phenotype "race-specific resistance" (Table 6). This means that the presence of the resistance gene R_1 in the plant is a pre-requisite for identifying the gene Avr_1 in the pathogen and vice versa. The genes for resistance and avirulence are conditional, their phenotypes are expressed only under special conditions, namely when the corresponding gene in the partner organism is present.

The gene-for-gene hypothesis for triggering race-specific resistance may be formulated employing terms commonly used in plant pathology: The interaction between a host plant and its pathogen results in incompatibility, implying that the pathogen is resisted by the plant only if a corresponding gene pair R_1 – Avr_1 (i.e., one in which the genes "recognize" each other) is present. By contrast, compatibility between both partners emerges, and the plant is parasitized, if the attacking pathogen is *not* recognized by its host plant because no alleles are present to form a corresponding gene pair R_1 – Avr_1 (see Table 6). The product of the pathogen's avirulence gene seems to serve as a "label" for its recognition by the plant's resistance gene. However, the host plant must have acquired this new genetic determinant R_1 sometime before, for example by spontaneous mutation. In other words, only the presence of the previously acquired genetic determinant in the host plant permits formation of the corresponding gene pair R_1 – avr_1 and hence the expression of race-specific resistance.

Table 6. Quadratic check showing how the expression of race-specific resistance depends on the genetic constitution of the host plant and pathogen

	Host plants (diploid)	
	susceptible $\frac{r_1}{r_1}$ (No resistance factor)	resistant $\frac{R_1}{R_1}$ (**Resistance factor**)
Pathogens (dikaryotic) $\frac{avr_1}{avr_1}$ (No avirulence factor)	Susceptibility (parasitism, compatibility)	Susceptibility (parasitism, compatibility)
(**Avirulence factor**) $\frac{Avr_1}{Avr_1}$	Susceptibility (parasitism, compatibility)	**Resistance** (**defense, incompatibility**)

Results obtained after infecting diploid plants with dikaryotic pathogens. The plants are either race-specific resistant, carrying genotype R_1/R_1 expressing a "resistance factor", or susceptible r_1/r_1 to infection. The pathogens are either avirulent, genotype Avr_1/Avr_1 expressing an "avirulence factor", or specific virulent, avr_1/avr_1. The phenotype of a host-pathogen interaction leading to parasitism, i.e., a virulent infection is called a compatible response, whereas the phenotype of an interaction that repells the pathogen, i.e., an avirulent infection, is called an incompatible response.

Pathogens carrying the avirulence gene Avr_1 are recognized and resisted by plants harboring the resistance gene R_1. By contrast, the same R_1 resistant plants may be parasitized by pathogens carrying the mutant allele avr_1. Such a mutant pathogen is unable to synthesize the avirulence factor A_1 recognized by the plant's race-specific resistance factor $Rcpt_1$. Thus the attacking pathogen escapes recognition and no defense reactions are triggered. Therefore, plants carrying resistance gene R_1 can be used to select, within a population of Avr_1 pathogens, single variants harboring the defective avr_1 allele. The ensuing virulent behavior of such mutant pathogens against plants carrying resistance gene R_1 has been called "specific virulence" and the avr_1 allele responsible for it a specific virulence allele. The term specific virulence was coined because the corresponding pathogen mutant specifically negates race-specific resistance determined by the plant's resistance gene R_1. However, expression of specific virulence determined by gene pair R_1 - avr_1 is suppressed in an $R_1 R_2$ resistant plant if a second corresponding gene pair R_2 - Avr_2 is present. This means that avirulence is epistatic to virulence. The term epistatic means that one gene interferes with the expression of another nonallelic gene. Thus the expression of specific virulence determined by the gene pair R_1 - avr_1 is suppressed by the avirulent response triggered by the second gene pair R_2 - Avr_2. Correspondingly, a haploid pathogen with the genetic constitution $avr_1 Avr_2$ (or a diploid or dikaryotic pathogen of genetic constitution $avr_1/avr_1 Avr_2/Avr_2$) does not express specific virulence determined by the pair R_1 - avr_1 since incompatibility caused by the second gene pair R_2 - Avr_2 suppresses the specific virulence determined by gene pair R_1 - avr_1. Only if the plant lacks the resistance gene R_2 and hence no gene pair R_2 - Avr_2 is formed, is specific virulence, determined by the pair R_1 - avr_1, expressed. In summary, a pathogen carrying a particular defective avirulence gene, for example, the specific virulence allele avr_a, can be identified and selected only on differentials or tester plants carrying the race-specific resistance gene R_a. All other pathogens which harbor the wild type allele Avr_a trigger incompatibility on the R_a resistant plants, thus preventing pathogen reproduction.

The presence or absence of a particular avirulence gene or avirulence factor may be shown in two ways: (1) By demonstrating an incompatible interaction like an HR on a corresponding race-specific resistant host plant. (2) By selecting on the race specific resistant plant those pathogens which lack the corresponding avirulence factor. Mutations for specific virulence map, except in rare cases (Chap. 9, Sect.9.1.8), within the avirulence gene whose expression they prevent. The analysis of segregating pathogen progenies on a set of host plant differentials, each carrying a single gene for race-specific resistance, provides an elegant method for selecting specific virulent segregants and mapping their genetic determinants.- In a similar way different race-specific resistance genes can be identified by infecting plants with each one of a set of specific virulent mutants.

When analyzing the genetic determination of avirulence Flor occasionally discovered deviations from a strict gene-for-gene relationship, suggesting that more than one gene in the pathogen might be involved in determining avirulence. Such deviations became apparent after crossing different pathogen races and inspecting the segregation among the progeny in the F_2 generation. If only one gene determines avirulence, a ratio of 3 : 1 avirulent to virulent is expected among the progeny (see *Fig. 3*). Instead, small but significant deviations from this ratio were found suggesting that more than one gene was involved in the determination of avirulence. Subsequently, Flor and others searched directly for deviations from the strict gene-for-gene relationships. However, in these experiments specific virulence was employed as a selectable allele for identifying genes involved in the expression of avirulence. Indeed, deviations were discovered not only in *M. lini* but also in other host plant/pathogen systems, e.g., *Puccinia graminis* f.sp. *tritici* and f.sp. *avenae* as well as *P. recondita* f.sp. *tritici* on cereals. Further examples demonstrating deviations from a strict gene-for-gene relationship in race-specific resistance are discussed in Chap. 9, Sect. 9.1.8.

Interactions between host plants and pathogens carrying alleles R_x or r_x and Avr_x or avr_x, respectively, governed by the rules of the gene-for-gene hypothesis may be formally described as follows (*Fig. 4a*): Recognition occurs between two substances, the plant's resistance product Rcpt, coded by resistance gene R, and the pathogen's avirulence product A, coded by the avirulence gene *Avr*. This triggers in the plant the expression of defense reactions, such as the hypersensitive reaction. However, if one or both genes and gene products determining recognition are lacking, or have become defective as a result of mutation, no defense reactions are triggered, and the pathogen is able to parasitize its host (*Fig. 4b*).

It is worth mentioning that a pathogen that has mutated to avr_1, i.e., to specific virulence, may appear also as a mutant with an extended "host range" when infecting a host plant carrying the gene R_1 for race-specific resistance. Such a situation could arise when a pathogen can overcome the basic resistance of two different plant species but is prevented from parasitizing one of them because it carries an unrecognized race-specific resistance gene R_x. If the corresponding Avr_x gene of that pathogen mutates to avr_x the R_x resistant plant can also be parasitized, resulting in an extension in the host range of that pathogen (see also Chap. 9, Sect. 9.1.3.1).

As mentioned above, there may be more than one pair of corresponding resistance and avirulence genes present in pathogen and host plant effecting the expression of race-specific resistance. Several different gene pairs may come into action if the corresponding resistance genes are present in the plant. In the course of time, many different avirulence genes have been identified in pathogens that correspond to several distinct resistance loci in the plant. Different individuals within a plant population may carry different resistance alleles mapping at the same chromosomal locus, each allele determining a particular resistance specificity and recognizing a particular avirulence product (see also Sect. 9.1.4, paragraph (4)). Crosses between plants carrying distinct resistance alleles at the same locus did not form recombinants. Neither susceptible plants nor plants with both

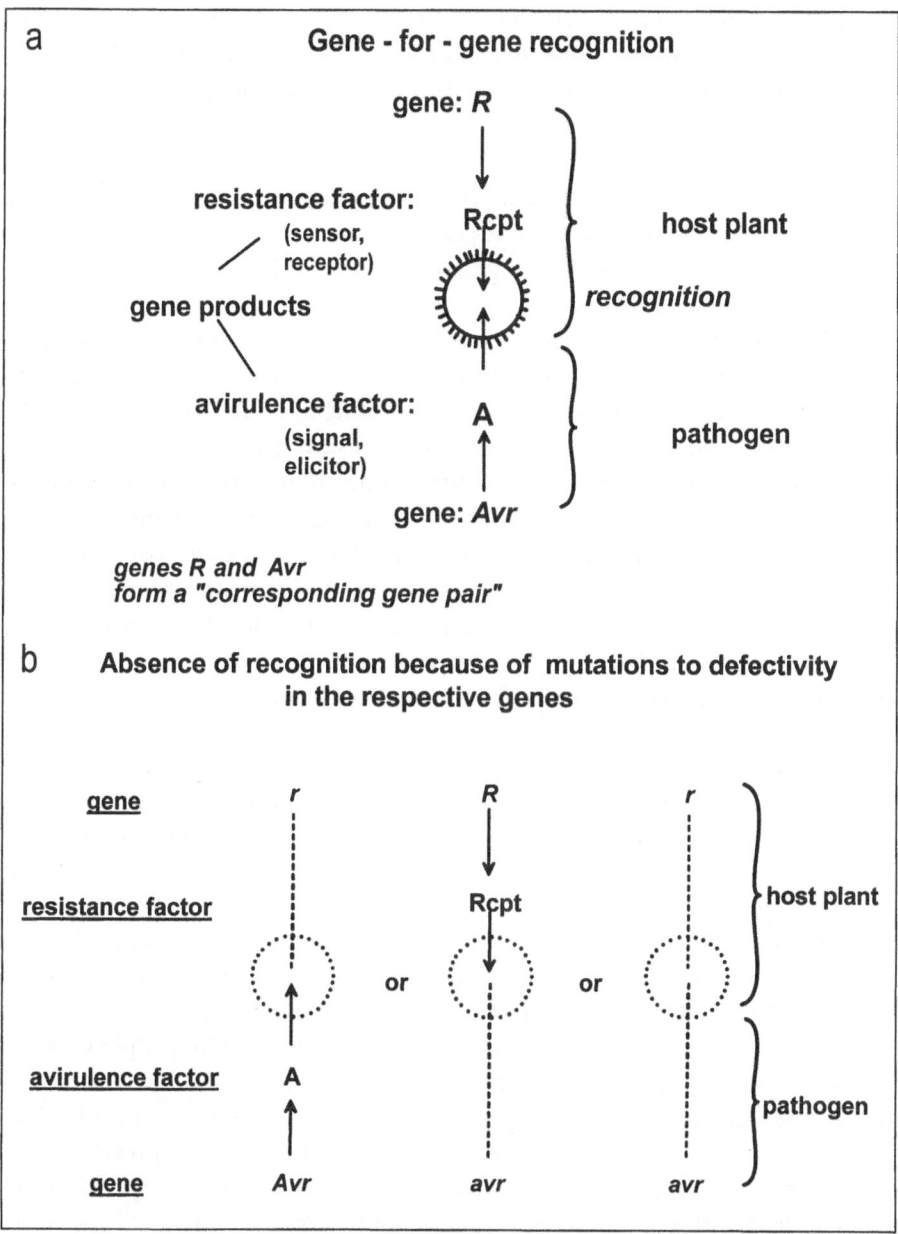

Figure 4a,b. The gene-for-gene interaction involved in recognition (or its failure) between host plant and pathogen, via their *R* and *Avr* genes, leading either to incompatible or to compatible interaction. **a:** Recognition takes place if both the resistance product Rcpt (the sensor), coded for by the race-specific resistance gene *R* of the host plant, and the avirulence product A (the signal) of the pathogen, coded for by its avirulence gene *Avr*, recognize each other. The event of recognition sets in motion a signal transduction chain finally releasing expression of defense genes effecting incompatibility. **b:** Recognition between the gene productes of plant and pathogen fails because either only one or both gene products are inactive. The defective avirulence product, or a complete lack of a product (symbolized by the *broken line*) causes the phenotype of specific virulence resistance specificities were obtained. To the extent that the progenies were large enough to detect rare recombinants this finding indicates that the different resistance specificities are not coded by closely linked resistance genes. The appearance of different resistance alleles at the same chromosomal locus seems to depend on processes involving unequal crossing over and so-called illegitimate recombination events within the same gene. The phenomenon of more than one resistance specificity coded for at one genetic locus is called multiple allelism for resistance (see also Sect. 9.1.4, paragraph (4)). It is an important factor in contributing to genetic polymorphism for race-specific resistance. Furthermore, no mutual influence or cross-reactions have been found between genes or their products belonging to different corresponding genes in plants or in pathogens.

Finally, incompatibility was found to be epistatic to compatibility and specific virulence. A single gene pair, like $R_1 - Avr_1$, able to trigger incompatibility, prevents expression of specific virulence effected by an inactive, corresponding allele pair like e.g. $R_2 - avr_2$. The three most prominent properties in the expression of race-specific resistance are summarized in Table 7: (1) a single corresponding gene pair ($R_1 - Avr_1$ or $R_2 - Avr_2$) is sufficient to trigger pathogen defense and incompatibility, respectively; (2) there is no cross-reaction between resistance and avirulence genes belonging to different gene pairs; (3) incompatibility is epistatic to compatibility and specific virulence, for example, $R_1 - Avr_1$ is epistatic to $R_2 - avr_2$, and $R_2 - Avr_2$ to $R_1 - avr_1$.

Necrotic hypersensitive defense reactions may be released by many different corresponding gene pairs. As a result the morphology of the necrotic lesions that appear may vary with the resistance gene concerned. Hence under favorable conditions a skilled plant pathologist can discriminate between different resistance genes triggering a particular defense reaction.

An interesting method of searching for likely gene-for-gene relationships that release cultivar-specific resistance without requiring crossing experiments has been based on infection experiments among different plant cultivars and pathogen races. The method is particularly useful because many fungal phytopathogens lack a sexual cycle, and the genetic recombination that would allow estimation of the number of genes in the pathogen that determine race-specific resistance. In 1959, Person demonstrated, in a theoretical analysis, that incompatible and compatible responses resulting from infecting different race-specific resistant cultivars with various pathogen races can be arranged into an "ideal quadratic check" if host plants and pathogens follow a strict gene-for-gene interaction. If the host-pathogen system is comprised of sufficient numbers of both differentials containing a single gene for race-specific resistance and distinct pathogen races collected in nature, one can determine whether the recorded incompatible and compatible responses fit the matrix predicted by the ideal quadratic check. Where the set of cultivars and pathogen races available is incomplete and corresponding pathogen races do not exist for all differentials, it may still be possible to find out if the emerging "gapped" pattern nonetheless corresponds to a strict

Table 7. Different race-specific resistance genes and corresponding gene pairs interact independently when expressing incompatibility. Incompatibility is epistatic over compatibility

	Host plants (diploid)		
	$\dfrac{R_1}{R_1}$	$\dfrac{R_2}{R_2}$	$\dfrac{R_1}{R_1}\ \dfrac{R_2}{R_2}$
Pathogens (dikaryotic)			
$\dfrac{Avr_1}{Avr_1}$	i	c	i
$\dfrac{avr_1}{avr_1}$	c	c	c
$\dfrac{Avr_2}{Avr_2}$	c	i	i
$\dfrac{avr_2}{avr_2}$	c	c	c
$\dfrac{Avr_1\ avr_2}{Avr_1\ avr_2}$	i	c	i
$\dfrac{avr_1\ Avr_2}{avr_1\ Avr_2}$	c	i	i

The table demonstrates in a quadratic check two characteristics of race-specific resistance: (1) Only the corresponding allele pairs R_1 – Avr_1 or R_2 – Avr_2 are able to release race-specific resistance, whereas alleles belonging to different pairs, such as R_1 and Avr_2 or R_2 and Avr_1, show no cross reaction (see responses within the double framed part of the table). (2) Inncompatibility is epistatic to compatibility: R_1 – Avr_1 over R_2 – avr_2 and R_2 – Avr_2 over R_1 – avr_1 (see reactions outside the double framed part of the table). – Resistance to an attacking pathogen is designated in the table as "i" (incompatibility), susceptibility or parasitism as "c" (compatibility).

gene-for-gene interaction. An example of inferring a gene-for-gene relationship from this kind of analysis is *Phytophthora infestans* causing late blight on the potato *Solanum tuberosum* because at the time of the analysis crosses among different blight races were not feasible. Person's kind of analysis is possible for the majority of host-pathogen systems in which sexuality is lacking but for which sufficient numbers of differentials and pathogen races are available. Person confirmed the validity of his method by showing that in some rare cases in which crossing experiments among pathogens had already excluded a gene-for-gene relationship obvious deviations from an ideal quadratic check could also be found. Deviations from the ideal quadratic check were also observed when one of the differentials was subsequently found to harbor two resistance genes.

However, there are limits to Person's method that must also be borne in mind: A quadratic check reflects the appearance of phenotypes that depend on the presence or absence of two determinative factors, one in the plant and one in the pathogen. Projecting this relationship onto a signal-sensor system, the expression of race-specific resistance depends, on the pathogen side, on one gene and its product serving as a signal and, on the plant side, on one gene and its product being able or unable to recognize the pathogen's signal. However, this assumes that the signal, i.e., the avirulence factor, after recognition by the resistance gene product, directly triggers expression of defense reactions. In other words, no additional gene activities are required for releasing the expression of defense reactions. The results of Flor's classic crossing experiments to explore the genetic determination of rust resistance in flax suggested such a direct connection. However, Flor also noticed exceptions that did not fit a simple gene-for-gene pattern for triggering expression of race-specific resistance (see Chap. 9, Sect. 9.1.8). The experimental results that could be interpreted as evidence of signal transduction events, forming a pathway between signal recognition and the release of defense reactions, are discussed in Chap. 9, Sect. 9.1.4, paragraph (4).

Even if infection experiments between different cultivars and pathogen races conform to Person's matrix this does not exclude the possibility that the interactions could be more complex and that no strict gene-for-gene relationship actually exists. For example, closely linked genes may be involved which can not be resolved by applying standard crossing techniques. Some examples of this are discussed in detail in Chap. 9, Sect. 9.1.8. Only genetic analysis carried out by crossing different host plant cultivars as well as pathogen races can ultimately substantiate the existence of a strict gene-for-gene interaction.

The observation that many genes involved in race-specific resistance are dominant is of special interest for basic research as well as for resistance breeding. All higher plants are diploid or polyploid, whereas fungal pathogens may be either haploid, such as *Ascomycetes* (e.g., *Magnaporthe grisea*, a pathogen on rice), diploid, such as *Oomycetes* (e.g., *Phytophthora infestans* on potato), or dikaryotic, such as the rust fungi belonging to *Basidiomycetes* (e.g., *Melampsora lini* on flax). In terms of their hereditary status dikaryotic and diploid fungi may be either homozygous or heterozygous, and in the latter case dominant genes are expressed and recessive genes are not. Recessive genes of dikaryotic or diploid organisms are expressed only if they are in an homozygous state.

The results of Flor's hybridization experiments showed that both the genes determining race-specific resistance in the flax host plant, *R*, and avirulence in the rust pathogen, *Avr*, are expressed dominantly or codominantly, whereas the determinants for infection susceptibility, *r*, and for defective avirulence and specific virulence, *avr*, respectively, are recessive (see also *Fig. 3*, segregation pattern in F_1 and F_2 generations). The expression of race-specific resistance, with alleles *R* or *r* either homozygous or heterozygous, is shown in the quadratic check of Table 8. The upper part of the table shows the interaction of different host plants with haploid pathogens, the lower part shows diploid or dikaryotic pathogens. The results reveal that in heterozygotes or heterokaryons one copy of each of the dominant genes R_1 and Avr_1 suffices for expression of race-specific resistance, whereas alleles r_1 and avr_1 behave as recessives.

Another tabular version of the strict gene-for-gene interaction, is the "reciprocal check", first used for *Puccinia graminis* f.sp. *tritici* on wheat: Each partner, host plant as well as pathogen, contains two genes of different specificity, both having a counterpart in the other partner and thereby forming a corresponding gene pair. However, each of the interacting partners carries both of these genes in reciprocal allele combinations, e.g., R_1/R_1 r_2/r_2 and r_1/r_1 R_2/R_2 for the plant and Avr_1/Avr_1 avr_2/avr_2 and avr_1/avr_1 Avr_2/Avr_2 for the pathogen. Table 9 demonstrates that there are no cross-interactions between components belonging to different gene pairs. If incompatible and compatible interactions between race-specific resistant cultivars and different pathogen races can be arranged in a quadratic check according to Table 6 and also in a double reciprocal check according to Table 9, this is recognized as the final proof for a strict gene-for-gene interaction as proposed by Flor.

There is a direct experimental indication that the resistant response in the host plant is triggered by recognition that occurs only during the infection process, strongly suggesting that recognition is mediated by particular surface structures. Employing *M. lini Avr₁* it was shown that the incompatibility of an R_1 resistant

Table 8. Dominance of genes determining race-specific resistance and avirulence

		Host plants (diploid)		
		$\dfrac{R_1}{R_1}$	$\dfrac{R_1}{r_1}$	$\dfrac{r_1}{r_1}$
Pathogens (haploid)	Avr_1	i	i	c
	avr_1	c	c	c
Pathogens (dikaryotic or diploid)	$\dfrac{Avr_1}{Avr_1}$	i	i	c
	$\dfrac{Avr_1}{avr_1}$	i	i	c
	$\dfrac{avr_1}{avr_1}$	c	c	c

Each gene of the corresponding pair $R_1 - Avr_1$ is dominant over its defective alleles r_1 and avr_1, respectively. In the table resistance, or an incompatible interaction, is indicated by "i", and susceptibility, or a compatible interaction, by "c". The upper part of the table shows infection of dipoid cultivars with haploid pathogens, the lower part with dikaryotic or diploid pathogens.

Table 9. Reciprocal check demonstrating independent expression of genes for race-specific resistance and incompatibility, respectively

	Host plants (diploid)	
	$\dfrac{R_1 \ r_2}{R_1 \ r_2}$	$\dfrac{r_1 \ R_2}{r_1 \ R_2}$
Pathogens (dikaryotic or diploid) $\dfrac{Avr_1 \ avr_2}{Avr_1 \ avr_2}$	**Resistance (defense, incompatibility)**	Susceptibility (parasitism, compatibility)
$\dfrac{avr_1 \ Avr_2}{avr_1 \ Avr_2}$	Susceptibility (parasitism, compatibility)	**Resistance (defense, incompatibility)**

The table shows results of interactions between two diploid cultivars and two diploid or dikaryotic pathogen races, each carrying reciprocal combinations of two different race-specific resistance alleles, $R_1/R_1 \ r_2/r_2$ and $r_1/r_1 \ R_2/R_2$, and corresponding avirulence alleles, $Avr_1/Avr_1 \ avr_2/avr_2$ and $avr_1/avr_1 \ Avr_2/Avr_2$. The genes are expressed independently of each other.

flax plant can be expressed only while infection by the pathogen proceeds. By contrast, no pathogen defense occurs in an R_1 resistant flax plant when a dikaryotic Avr_1/avr_1 aecidium is developing within it and aecidiospores are formed. The latter situation arises when a receptive hypha of a haploid mycelium carrying the specific virulence allele avr_1 borne on an R_1 resistant plant is fertilized by a pycnidiospore originating from a haploid wild type Avr_1 mycelium growing in a susceptible flax cultivar. Under these circumstances the pathogen's Avr_1 gene product obviously does not contact the recognition system coded by gene R_1 and therefore no defense reactions are triggered (Day 1972).

The dominant expression of both the resistance gene of the host plant and the avirulence gene of the pathogen corroborates the assumption of the preceding discussion that the R and Avr genes synthesize products that are responsible for the expression of pathogen defense, whereas the r and avr alleles either make defective gene products unable to trigger defense, or no corresponding gene products at all. This interpretation is supported, so far as the expression of resistance is concerned, by the observation that the degree of resistance in a *Triticum monococcum* cultivar produced by a single copy of gene $Sr22$ against *Puccinia graminis* f.sp. *tritici* indicated by the infection type, decreases with increasing ploidy, i.e., diploid, tetraploid, or hexaploid cultivars (Kerber and Dyk 1973).

Dominant expression of resistance by the plant and recessive expression of specific virulence by the pathogen have important consequences for diploid and dikaryotic pathogens: Many of these pathogens in natural populations may be heterozygous Avr_1/avr_1 and are avirulent on an R_1 race-specific resistant cultivar even if they also carry the recessive specific virulence allele avr_1. If such a pathogen happens to infect a susceptible cultivar, pass through a sexual cycle resulting in genetic recombination and formation of a homozygote avr_1/avr_1, a new pathogen type emerges which is (race-)specific virulent against the R_1 resistant plant. In this way new virulent pathogen races are generated by sexual recombination within the heterozygous rust population. The involvement of sexuality in the appearance of new pathogen races was convincingly demonstrated in the diploid pathogen *Phytophthora infestans* causing potato blight. This pathogen has two mating types, A_1 and A_2, and mating with eventual sexual recombination occurs between male and female sexual organs, antheridia and oogonia, of opposite mating types. In Europe only one mating type, A_2, exists, thus generally excluding sexual recombination. By contrast, in one particular region in Mexico both A_1 and A_2 mating types of *P. infestans* are present allowing for sexual recombination. Under these conditions, newly specific virulent races appear much more rapidly than in Europe. Thus heterozygosity may be an important "genetic reservoir" for the emergence of new pathogen physiological races via sexual recombination.

The quadratic check of Table 10 summarizes the outcome of interactions in all possible allele combinations between diploid host plants and diploid or dikaryotic phytopathogenic fungi, each carrying two copies of each gene for race-specific resistance and avirulence, respectively, either in wild type or mutant allelic form. The table demonstrates first the dominant and epistatic expression of corresponding gene pairs and second the absence of cross reactions between different corresponding gene pairs and their various alleles.

The term virulence requires some additional discussion because it is employed – unfortunately – in two completely different contexts. On the one hand, it is used to describe a phenomenon observed after infection of race-specific resistant plants by particular pathogen races. On the other hand, it is used in the context of toxin producing pathogens. As discussed above, the term specific virulence is used to describe the ability of an infecting biotrophic pathogen to negate the race-specific resistance gene of its host plant. Since there is no functional pathogen avirulence gene product the product of the resistance gene of the plant has nothing to recognize. Under these conditions the pathogen is virulent and colonizes the plant. In this case virulence is thus a qualitative phenomenon based on *non*-recognition between plant and pathogen. By contrast, the term virulence, as employed with toxin forming pathogens, characterizes the action of the toxin quantitatively: high virulence depends on a very effective toxin action, low virulence on less effective action. While the specific virulence of biotrophic pathogens is based on a loss of recognition specificity allowing the pathogen to parasitize its host, the virulence of necrotrophic pathogens describes quantitatively the action of toxins that produce disease symptoms that vary in severity. Hence, the term virulence is used to describe quite different phenomena.

Table 10. Dominance and epistasis of race-specific resistance and independence among different corresponding gene pairs

	Host plants								
	$\dfrac{R_1\ R_2}{R_1\ R_2}$	$\dfrac{R_1\ R_2}{r_1\ R_2}$	$\dfrac{r_1\ R_2}{r_1\ R_2}$	$\dfrac{R_1\ R_2}{R_1\ r_2}$	$\dfrac{R_1\ R_2}{r_1\ r_2}$	$\dfrac{r_1\ R_2}{r_1\ r_2}$	$\dfrac{R_1\ r_2}{R_1\ r_2}$	$\dfrac{R_1\ r_2}{r_1\ r_2}$	$\dfrac{r_1\ r_2}{r_1\ r_2}$
Pathogens									
$\dfrac{Avr_1\ Avr_2}{Avr_1\ Avr_2}$	i	i	i	i	i	i	i	i	c
$\dfrac{Avr_1\ Avr_2}{avr_1\ Avr_2}$	i	i	i	i	i	i	i	i	c
$\dfrac{avr_1\ AVr_2}{avr_1\ Avr_2}$	i	i	i	i	i	i	c	c	c
$\dfrac{Avr_1\ Avr_2}{Avr_1\ avr_2}$	i	i	i	i	i	i	i	i	c
$\dfrac{Avr_1\ Avr_2}{avr_1\ avr_2}$	i	i	i	i	i	i	i	i	c
$\dfrac{avr_1\ Avr_2}{avr_1\ avr_2}$	i	i	i	i	i	i	c	c	c
$\dfrac{Avr_1\ avr_2}{Avr_1\ avr_2}$	i	i	c	i	i	c	i	i	c
$\dfrac{Avr_1\ avr_2}{avr_1\ avr_2}$	i	i	c	i	i	c	i	i	c
$\dfrac{avr_1\ avr_2}{avr_1\ avr_2}$	c	c	c	c	c	c	c	c	c

The table shows the effect of various allele combinations of two different corresponding resistance and avirulence genes each one in an active or inactive form (R_1, r_1, R_2, r_2 and Avr_1, avr_1, Avr_2, avr_2). Notice in this quadratic check: (1) the dominance of R and Avr over r and avr alleles as well as epistasy of incompatibility over compatibility of different gene pairs (e.g., in the pathogen-plant interaction $Avr_1/avr_1\ avr_2/avr_2 - R_1/r_1\ R_2/r_2$) and (2) the absence of cross reactions between different corresponding gene pairs, irrespective of the alleles the gene pair may carry. Host plants and pathogenic fungi are diploid or dikaryotic, respectively. "i" designates incompatible, "c" compatible interaction.

It is worth noting that interactions between plants and phytopathogenic fungi launched by signal-sensor recognition reactions may lead to quite different results: Recognition of host-specific toxins produced by necrotrophic pathogens results in the death and parasitism of the afflicted host plant, whereas recognition of the elicitor of biotrophic pathogens by race-specific resistant plants effects defense against the attacking pathogen preventing parasitism of the host plant. Despite these opposite results – parasitism or its prevention – recognition has the same result at the cell level in both cases, namely, the death of the attacked plant cell. Killing by necrotrophs facilitates plant colonization since the dead cells are a source of food for the parasite, whereas biotrophic pathogens are prevented from colonizing and parasitizing the plant by triggering a hypersensitive defense reaction.

In the past the question has been raised of whether active basic resistance could be triggered like race-specific resistance by specific gene-for-gene interactions. However, this would require in the plant a huge "reserve" of specific receptors able to recognize an enormous diversity of specific pathogen-synthesized elicitors (Heath 1991). But then how could plants have acquired, and preserved during evolution, reserves of specific receptors for elicitors they had either never met before or had met only very recently? Furthermore, how could one explain the startling similarity between the defense mechanisms of active basic resistance to pathogen attack on the one hand, and the responses to abiotic stresses such as wounding, or injury by chemicals, on the other? If recognition between non-host plants and pathogens is an integral part of active basic resistance as it is in race-specific resistance, as in the gene-for-gene interaction, one should expect that single mutational changes in the pathogen's elicitor molecule would abolish recognition and effect basic compatibility for that particular pathogen. However, so far no single-step mutation has been observed that resulted in the emergence of a new *forma specialis* of a pathogen at frequencies comparable to those observed in the appearance of new, specific virulent pathogen mutants. Hence it is highly unlikely that a mechanism identical to that triggering race-specific resistance could also release basic resistance, namely the interaction of a pathogen-borne elicitor and a plant resident receptor. Clearly, this statement does not exclude the possibility that gene-for-gene interactions, or other signal-sensor reactions, are involved in triggering active basic resistance (see Chap. 6). Also, gene-for-gene interactions seem to be involved in heterologous hypersensitive reactions (see Chap. 9, Sect. 9.1.3.1) and in this way they may also determine the host range of a pathogen i.e., the *forma specialis* (Heath 1991). Active basic resistance differs from race-specific resistance mainly because of its much more complex nature: It includes many different mechanisms precluding colonization by a multitude of very different pathogens, and the defense mechanisms may depend on the plant, the pathogen, or both.

Reviews

Brettell R.I.S., Pryor A.J. (1986): Molecular approaches to plant and pathogen genes. In: Blonstein A.D., King P.J. (eds.): Plant Gene Research: A Genetic Approach to Plant Biochemistry. Springer-Verlag, Vienna, New York. 233–246

Burnett J.H. (1975): Chapter 13: General aspects of fungal pathogenicity. In: Burnett J.H. (ed.): Mycogenetics. John Wiley & Sons, London, New York, Sidney, Toronto. 259–287

Crute I.R. (1985): The genetic bases of relationships between microbial parasites and their hosts. In: Fraser R.S.S. (ed.): Mechanisms of Resistance to Plant Diseases. Martinus Nijhoff/Dr.W.Junk, Dordrecht, Boston, Lancaster. 81–142

Crute I.R., de Wit P.J.G.M., Wade M. (1985): Mechanisms by which genetically controlled resistance and virulence influence host colonization by fungal and bacterial parasites. In: Fraser R.S.S. (ed.): Mechanisms of Resistance to Plant Diseases. Martinus Nijhoff/Dr.W.Junk, Dordrecht, Boston, Lancaster. 197–309

Day P.R. (1972): The genetics of rust fungi. In: Bingham R.T., Hoff R.J., McDonald G.I. (eds.): Biology of rust resistance in forest trees. U.S. Department of Agriculture, Forest Service, Washington, D.C. 3–17

Dinoor A., Eshed N., Nof E. (1988): *Puccinia coronata*, crown rust of oat and grasses. Advances in Plant Pathology 6: 333–344

Dixon R.A., Lamb C.J. (1990): Molecular communication in interactions between plants and microbial pathogens. Annu.Rev.Plant Physiol. **41**: 339–367

Ellingboe A.H. (1976): Genetics of host-parasite interactions. In: Heitefuss R., Williams P.H. (eds.): Encyclopedia of Plant Physiology, (NS), Physiological Plant Pathology. Springer Verlag, Heidelberg. 761–778

Flor H.H. (1954): Identification of races of flax rust by lines of single rust-conditioning genes. U.S.D.A. Techn. Bull. **1087**: 1–25

Flor H.H. (1956): The complementary genic systems in flax and flax rust. Adv.Genetics 8: 29–54

Flor H.H. (1971): Current status of the gene-for-gene concept. Annu.Rev.Phytopathol. **9**: 275–296

Gabriel D.W., Rolfe B.G. (1990): Working models of specific recognition in plant-microbe interactions. Annu.Rev.Phytopathol. **28**: 365-391

Heath M.C. (1991): Evolution of resistance to fungal parasitism in natural ecosystems. New Phytol. **119**: 331–343

Johnson R. (1992): Past, present and future opportunities in breeding for disease resistance, with examples from wheat. Euphytica **63**: 3–22

Johnson R. (1992): Reflections of a plant pathologist on breeding for disease resistance, with emphasis on yellow rust and eyespot of wheat. Plant Pathology **41**: 238–254

Johnson R., Knott D.R. (1992): Specificity in gene-for-gene interactions between plants and pathogens. Plant Pathology **41**: 1–4

Keen N.T. (1982): Mechanisms conferring specific recognition in gene-for-gene plant parasite systems. In: Wood R.K.S. (ed.): Active Defence Mechanisms in Plants. Plenum Press, New York, London. 67–84

Keen N.T. (1982): Specific recognition in gene-for-gene host-parasite systems. Advances in Plant Pathology **1**: 35–81

Keen N.T. (1990): Gene-for-gene complementarity in plant-pathogen interactions. Annual Review of Genetics **24**: 447–463

Keen N.T. (1992): The molecular biology of disease resistance. Plant Mol.Biol. **19**: 109–122

Keen N.T. (1993): An overview of active disease defense in plants. In: Fritig B., Legrand M. (eds.): Mechanisms of Plant Defense Responses. Kluwer Academic Publishers, The Netherlands. 3–11

Knogge W., Hahn M., Lehnackers H., Rüpping E., Wevelsiep L. (1991): Fungal signals involved in the specificity of the interaction between barley and *Rhynchosporium secalis*. In: Hennecke H., Verma D.P.S. (eds.): Advances in Molecular Genetics of Plant-Microbe Interactions. Dordrecht, Kluwer Academic Publishers, Netherlands. 250–253

Lawrence G.J. (1988): *Melampsora lini*, rust of flax and linseed. Advances in Plant Pathology 6: 313–331

Littlefield L.J., Heath M.C. (1979): Ultrastructure of Rust Fungi. Academic Press, New York, San Francisco, London

Mansfield J., Bennett M., Bestwick C., Woods-Tör A. (1997): Phenotypic expression of gene-for-gene interaction involving fungal and bacterial pathogens: Variation from recognition to response. In: Crute I.R., Holub E.B., Burdon J.J. (eds.): The Gene-for-Gene Relationship in Plant-Parasitic Interactions. CAB International, Oxon UK, New York NY. 265–291

Person C. (1959): Gene-for-gene relationships in host : parasite systems. Can.J.Bot. 37: 1101–1130

Pryor T. (1987): The origin and structure of fungal disease resistance genes in plants. Trends Genet. 3: 157–161

Shaner G., Stromberg E.L., Lacy G.H., Barker K.R., Pirone T.P. (1992): Nomenclature and concepts of pathogenicity and virulence. Annu.Rev.Phytopathol. 30: 47–66

Scholtens-Thoma I.M.J., Joosten M.H.A.J., de Wit P.J.G.M. (1991): Appearance of pathogen-related proteins in plant hosts – Relationships between compatible and incompatible interactions. In: Cole G.T., Hoch H.C. (eds.): The Fungal Spore and Disease Initiation in Plants and Animals. Plenum Press, New York, London. 247–265

Thompson J.N., Burdon J.J. (1992): Gene-for-gene coevolution between plants and parasites. Nature 360: 121–125

Relevant papers

Flor H.H. (1942): Inheritance of Pathogenicity in *Melampsora lini*. Phytopathology 32: 653–669

Flor H.H. (1946): Genetics of pathogenicity in *Melampsora lini*. Jour.Agr.Res. 73: 335–366

Flor H.H. (1955): Host-parasite interaction in flax rust – its genetics and other implications. Phytopathogogy 45: 680–685

Heath M.C. (1991): The role of gene-for-gene interactions in the determination of host species specificity. Phytopathology 81: 127–130

1911 Kerber E.R., Dyck P.L. (1973): Inheritance of stem rust resistance transferred from diploid wheat (*Triticum monococcum*) to tetraploid and hexaploid wheat and chromosome location of the gene involved. Can.J.Genet.Cytol. 15: 397–409

9.1.2
The Properties of Race-Specific Resistance

In the following section, seven points summarizing the essential properties of race-specific resistance and the underlying gene-for-gene hypothesis are presented:

(1) A gene for race-specific resistance, R, carried by a host plant can be expressed only if the infecting pathogen harbors a gene Avr for avirulence, hence the presence of a corresponding gene pair R_1 – Avr_1 or R_x – Avr_x is required. However, if the corresponding gene carried, for example, by a haploid pathogen is mutant avr_1 instead of wild type Avr_1 no resistance is expressed. Genes for race-specific resistance in the plant, and for avirulence in the pathogen, are expressed conditionally, since each gene is expressed only in the presence of the corresponding partner gene. Furthermore, a plant-pathogen interaction R_x – Avr_x effecting avirulence and resistance, is epistatic over other

interactions leading to virulence and susceptibility, such as interactions of the type $R_x - avr_x$, $r_x - Avr_x$ or $r_x - avr_x$.

(2) Pathogens that carry specific virulence alleles such as avr_1 may be selectively multiplied on particular host plants which carry only the corresponding race-specific resistance gene R_1. Other pathogen races containing the wild type gene Avr_1 are excluded by the gene R_1. Thus, specific virulence alleles of a particular avirulence gene may be selected on a host plant carrying the corresponding race-specific resistance gene.

(3) R genes and Avr genes are expressed dominantly or codominantly in hetero-zygotes and each one synthesizes a gene product. By contrast, susceptibility and specific virulence are recessive traits in which corresponding gene products are either defective or absent.

(4) Expression of race-specific resistance may be triggered by each of many different corresponding gene pairs. Several plant resistance genes show genetic polymorphism, or multiple alleles. Furthermore different corresponding gene or allele pairs may trigger morphologically different hypersensitive reactions. – Cross reactions between the products of different corresponding R – Avr gene pairs do not occur.

(5) The correspondence between avirulence and race-specific resistance genes, the gene-for-gene relationship, may be explained phenotypically by the following model: A specifically acting avirulence gene product, A, coded by the pathogen's gene, Avr, is recognized by a specific resistance gene product, Rcpt, coded by the plant's corresponding resistance gene, R. This recognition constitutes the basic requirement for triggering race-specific defense reactions against pathogen attack. A single-step mutation in either R or Avr, causing a defect in or the absence of the corresponding gene product, prevents recognition and the expression of defense reactions.

(6) The large scale employment of cultivars with novel race-specific resistance genes in agriculture led, over time, to the selection of pathogens carrying novel types of specific virulence. In this way a coevolution of resistant host plants and specific virulent pathogens came into being which would not have occurred without human intervention (see also Chap. Sects. 9.1.5, 9.1.7).

(7) The recognition events involved in both active basic and race-specific resistance trigger in plants the expression of reactions belonging to a common battery of defense mechanisms (see also Fig. 5).

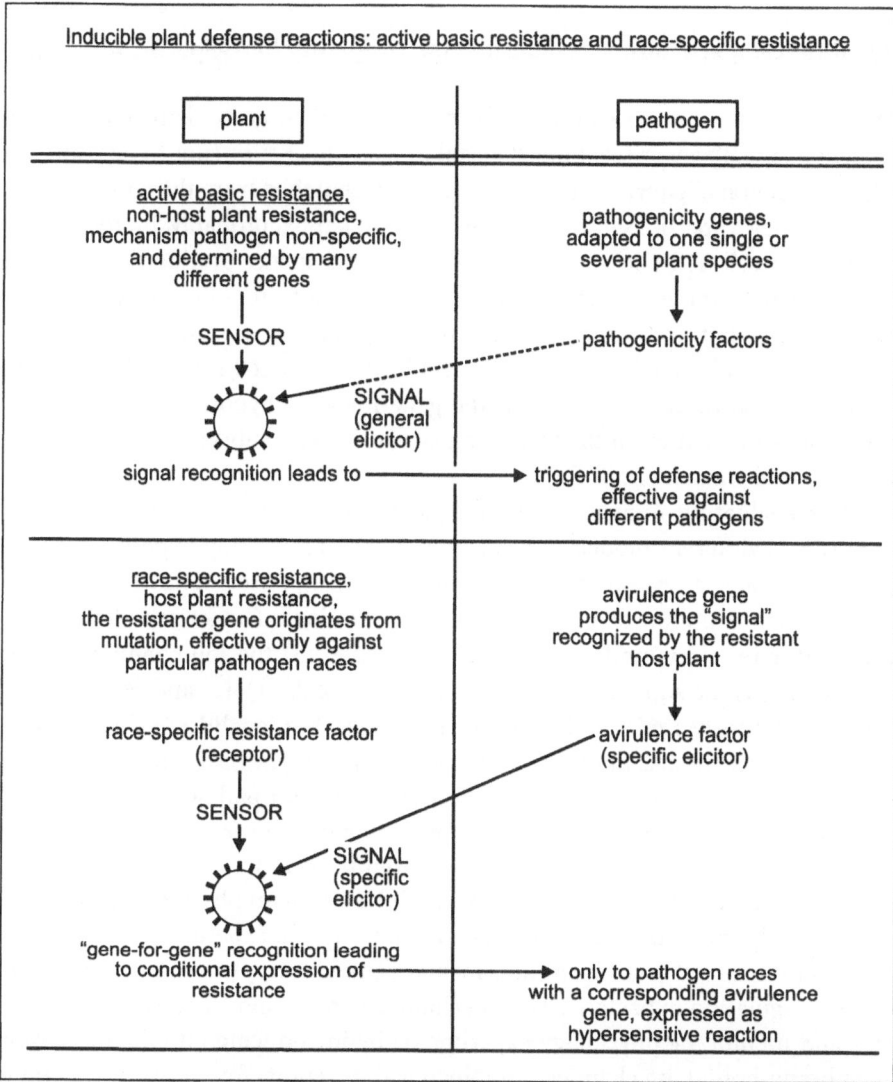

Figure 5. Schematic representation of active basic resistance and race-specific resistance. Both depend on possibly different signal-sensor reactions and different signal transduction chains. The principle in releasing active basic resistance is shown in the *upper* part of the figure, the principle in releasing race specific resistance in the *lower*. The release of race-specific resistance depends on overcoming basic resistance, i.e., on existing basic compatibility. Both active basic resistance as well as race-specific resistance employ the same battery of plant defense mechanisms

9.1.3
Hypothetical Mechanisms Responsible for Triggering Race-Specific Resistance

The recognition process summarized in paragraph (5) above is believed to represent a signal-sensor reaction: A signal or specific elicitor (not to be confused with the general elicitor, Chap. 6 and *Fig. 5*) is dispatched from the pathogen and is received by a specific sensor or receptor located, most probably, on the surface of the plant cell. The elicitor corresponds to the avirulence product, A, while the sensor or receptor is represented by the resistance factor, Rcpt (*Fig. 4a*). The ensuing recognition event between them generates a signal transduction pathway that ultimately affects the site(s) in the plant cell responsible for activating defense reactions. However, the gene-for-gene hypothesis does not address the actual nature of the processes, structures and substances participating in the signal transduction.

After evaluating numerous physiological, biochemical and genetic experiments, four different models were proposed. All are working hypotheses to explain race-specific resistance and they concern the supposed nature of the recognition reaction and the expression of defense reactions. The four models are: (1) the elicitor-receptor model, (2) the dimer model, (3) the ion channel defense model and (4) the suppressor-receptor model. Models (1), (2) and (4) were proposed in the early 1980s, and (3) in the late 1980s, but for didactical reasons we will discuss the models in the order of their numbers. The basis for all four models is the gene-for-gene relationship between host and pathogen for triggering race-specific resistance. However, each model is derived from different underlying concepts and from more or less differently conceived and/or interpreted experiments. Thus the elicitor-receptor model is based on physiological and biochemical experiments. The dimer model applies to the experiments of the elicitor-receptor model a stringent and formal genetic interpretation that refers to genetic regulation in bacteria. The ion channel defense model departs from electro- and membrane physiology experiments by introducing into the discussion membrane-bound ion channels combined with receptors for the elicitor, enzyme complexes, and second messengers which together form signal transduction chains that can alter the metabolic activities of the plant cell. Finally, the suppressor-receptor model refers to the same experimental results as the elicitor-receptor model but interprets them using different assumptions (Bushnell and Rowell 1981).

Reviews

Ellingboe A.H. (1976): Genetics of host-parasite interactions. In: Heitefuss R., Williams P.H. (eds.): Encyclopedia of Plant Physiology, (NS), Physiological Plant Pathology. Springer Verlag, Heidelberg. 761–778
Ellingboe A.H. (1981): Changing concepts in host-pathogen genetics. Annu.Rev.Phytopathol. **19**: 125–143

Ellingboe A.H. (1982): Genetical aspects of active defence. In: Wood R.S.K. (ed.): Active Defence Mechanisms in Plants. Plenum Press, New York. 179 – 192

Gabriel D.W., Rolfe B.G. (1990): Working models of specific recognition in plant-microbe interactions. Annu.Rev.Phytopathol. **28**: 365 – 391

Relevant papers

Bushnell W.R., Rowell J.B. (1981): Suppressors of defense reactions: A model for roles in specificity. Phytopathology **71**: 1012 – 1014

9.1.3.1
The Elicitor-Receptor Model

The elicitor-receptor hypothesis proposed to account for race-specific resistance in the presence of basic compatibility suggests that two groups of plant genes are involved: firstly the gene acting as sensor within the signal-sensor reaction that recognizes the pathogen, and secondly the several genes that express the plant's defense reactions. This hypothetical two-step process provides a plausible explanation of how the high specificity of race recognition always effects expression of several genes belonging to the same battery of plant defense genes. However, the elicitor-receptor hypothesis does not explain how recognition by the plant turns on expression of the plant defense genes.

According to the elicitor-receptor model, the release of race-specific resistance proceeds as follows: The pathogen's avirulence gene *Avr* either directly produces a signal, the elicitor, or the avirulence gene encodes an enzyme that produces an elicitor from pathogen material. The signal has also been called a "specific elicitor", because of its high selectivity in releasing defense against only particular pathogen races. Elicitors synthesized directly by the pathogen or which are break-down products of pathogen-borne material, are "exogenous elicitors". (If, as in active basic resistance, the break-down product is derived from the plant, one speaks of an endogenous elicitor). The plant's recognition of the exogenous elicitor of race-specific resistance, i.e., its sensor, receptor and resistance factor, respectively, is genetically determined by the race-specific resistance gene. Since each resistance factor recognizes only one corresponding avirulence factor and specific elicitor one speaks of a pair of corresponding resistance- and avirulence genes determining race-specific resistance. The plant's receptor is presumed to be a membrane protein localized in the plasma membrane. Because of this location the receptor stays in close contact with the cell wall, providing for the earliest possible meeting between elicitor and cell surface. Elicitor, receptor, and the released defense reactions are the salient elements in the elicitor-receptor model. These are discussed in detail below.

Elicitors: What is the chemical nature of the specific elicitor (*Fig. 5*, lower part) encoded by the avirulence gene *Avr*? Suppose that the avirulence factor is an exogenous elicitor synthesized by, or derived from, the pathogen, then

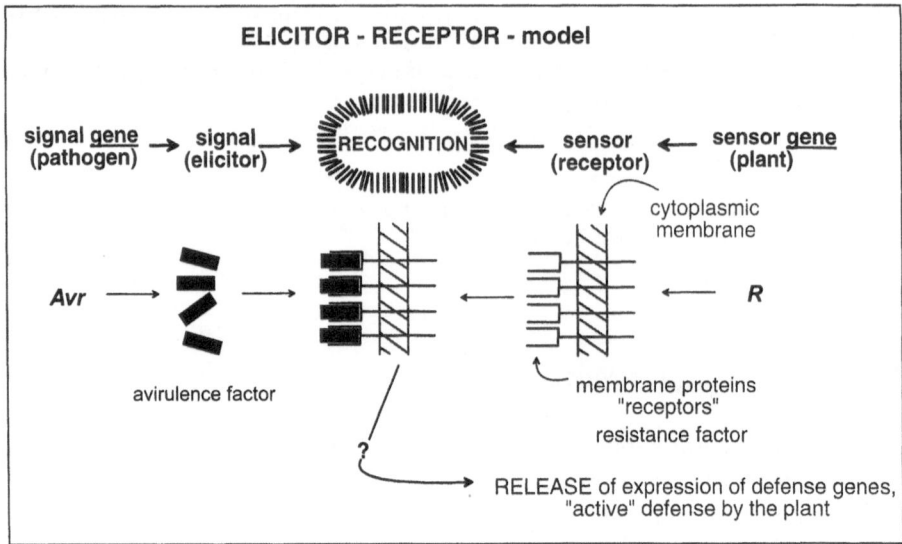

Figure 6. The elicitor-receptor model for gene-for-gene interaction: The *upper* part of the figure shows the hypothetical signal-sensor reaction. The *lower* part shows the components and processes of the model: (1) the avirulence factor, synthesized by avirulence gene *Avr*, represents a signal molecule, the elicitor, and (2) the receptor molecule synthesized by the host plant's resistance gene *R*, is the sensor. Recognition between signal and sensor consists of specific binding of the avirulence factor or elicitor to the receptor, releasing expression of the plant's defense reactions. For simplicity the elicitor is shown as directly synthesized by the pathogen. Further details are given in the text

its chemical nature might vary according to the metabolic versatility of the fungus. Elicitors could be primary or secondary fungal metabolites, they could be particular peptides or proteins derived from metabolites, or they could be breakdown products of carbohydrate polymers, e.g., a glucan oligosaccharide from the chitin cell wall of a pathogenic fungus. However, the specific elicitor could also be endogenous, i.e., of plant origin but arising from the pathogen's enzymatic activity. But this has so far only been demonstrated for elicitation of basic resistance and not for race-specific resistance.

The characterization of elicitors releasing expression of phytoalexins (see Chap. 5) was first carried out by Albersheim's group. From their experimental results they proposed the elicitor-receptor model for release of plant defense responses. Albersheim used the bacterial phytopathogen *Erwinia carotovora* (*Ec*) and germlings of its host plant soybean (*Glycine max*) as an experimental system for testing resistance responses (Davis et al. 1984, 1986). The endogenous elicitor, derived from plant cell walls, was later identified as an oligogalacturonide releasing synthesis of pterocarpan phytoalexins. A second system was developed by Hahlbrock and Scheel who employed potato slices, leaves, or potato cell suspension cultures to study the synthesis of the phytoalexin rishitin and related derivatives triggered by elicitors (Rohwer et al. 1987). The *host* plant system was potato (*Solanum tuberosum*) and *Phytophthora infestans* (*Pi*), and the *non-host* system potato with *P. megasperma* f.sp. *glycinea* (*Pmg*). The signal for eliciting resistance responses was an exogenous elicitor isolated from culture filtrates of both *Phytophthora* species. This elicitor was later identified as a glucan polysaccharide derived from the pathogen's cell walls. In experiments performed in both *host* systems, soybean-*Ec* and potato-*Pi*, in compatible and incompatible combinations (host-pathogen pairs with specific virulent and avirulent pathogens), it turned out that in each combination the added elicitor triggered phytoalexin synthesis. This occurred even in the *non-host* system potato-*Pmg*. Since the observed recognition reaction showed no race specificity it must be an active basic resistance response. To distinguish elicitors of active basic resistance from specific elicitors of race-specific resistance, the former are called "general elicitors" (*Fig. 5*, upper part), since they often cause defense reactions in several, often unrelated, plant species.

Although neither of the experimental signal-sensor systems involving *Phytophthora* species trigger expression of race-specific resistance, they mirror the general biochemical and molecular biological features of elicitor-receptor interactions. In the soybean *E. carotovora* system (Davis et al. 1984, 1986; Jin et al. 1984) an endopolygalacturonic acid lyase of the pathogen digests the pectin middle lamella of the plant cell wall. Among several digestion products, two oligo-alpha-1,4 D-galacturonides containing ten and eleven monomers are formed; these are the most effective endogenous elicitors so far found. Synthesis of phytoalexins may also be triggered by other comparable endogenous elicitors that result from wounding, or the action of UV-light, heavy metal ions, detergents or chloroform on plants. In all of these examples, elicitors are the product of plant-derived enzymes. By contrast, in the systems potato with *Pi* and *Pmg* (Rower et al. 1987, Fritzemeier et al. 1987), the exogenous elicitor is a heptaglucan with a particular branched structure, hepta-β-glucosidalditol, formed by digestion of the glucan polymer in the pathogen's cell wall. The enzyme responsible for producing this elicitor is a plant-derived β-glycosylhydrolase belonging to the defense related proteins. The heptaglucan is also an elicitor of defense reactions in several other plant species. Elicitors derived from digestion of cell walls of plants or pathogens exhibit neither specificity for one particular plant species nor are they active in every plant. However, as they never cause hypersensitive cell death, they are referred to as non-HR elicitors. These elicitors are frequently used in experimental model systems designed to analyze the reactions that follow elicitor-receptor interactions (see also Chap. 9, Sect. 9.1.4), since without host cell death the synthetic activities that follow the interactions can be studied.

The lack of race specificity observed in plant defense reactions triggered by the endogenous oligogalacturonide elicitors released by *E. carotovora* is understandable considering their origin. Enzymatic digestion of the rather highly uniform pectin polymer would not provide the many structurally diverse digestion products required for triggering expression of race-specific resistance. Of course, the same argument holds for the exogenous elicitor of *P. megasperma*, which originates from digestion of the glucan polymer of the pathogen's cell wall. Hence the detailed analysis by Albersheim is unlikely to answer questions about the nature of the specific elicitors that trigger race-specific resistance.

Plant cell receptors are supposed being located on the cell surface (*Fig. 6*), where they recognize arriving chemical signals and transduce the "messages" they represent inside the cell. In this way receptors form a contact between the environment and the cell's interior. The cell membrane receptor proteins which recognize specific elicitors are encoded by the plant's genes for race-specific resistance (*Fig. 6*).

The minimal functions assigned to receptors that recognize specific elicitors are: (1) extracellular binding of a ligand, the signal, which on a molecular level is the first step towards "recognition" between a specific elicitor and its receptor, (2) transmission of the signal generated to the internal membrane surface by binding between the ligand and its receptor, and (3) activation by the transmitted signal of a so-called effector. In this case the term effector covers all induced processes leading to expression of defense genes in the plant cell nucleus. However, the elicitor-receptor model specifies neither the structure of the corresponding receptors nor does it define the nature and mechanisms of action of the effectors. In summary, race-specific resistance triggered in this way manifests itself in metabolically active cells through the expression of active defense reactions induced by particular effectors.

The defense reactions expressed by race-specific resistance belong to the same battery of defense reactions employed in the active basic resistance of non-host plants. However, in the elicitor-receptor model the process of pathogen recognition, which triggers resistance responses by the host plant, is superimposed on the reaction chains that establish basic compatibility between plant and pathogen. This implies that up to and during the expression of race-specific resistance the pathogenicity genes of the infecting fungus are not turned off by the action of the resistant host plant (see Chap. 9).

Generally, two kinds of defense mechanisms can be found in plants defending themselves against pathogen attack. The first consists of immediate (within one or a few hours) cell death, so that the parasite loses its nutritive base. This is the hypersensitive reaction or hypersensitive cell death (HR). The second mechanism involves activation of genes so that new defense barriers are erected by the plant. The corresponding genes and their products contributing to plant defense, the pathogenesis related proteins, were already discussed in Chap. 5.

The hypersensitive reaction, HR, deserves special attention, because it is the most common defense mechanism in race-specific resistance against attack by biotrophic pathogens. HR results in rapid killing of the infected host cell and most of the neighboring cells, even though the latter were not penetrated by the pathogen. In contrast to cell death caused by senescence, the hypersensitive cell death is genetically programmed and its expression spread by signal transduction chains. Called programmed cell death (PCD) this killing of infected and some noninfected neighboring cells also leads to death of the parasite because it loses its nutritive base. Therefore, the formation of phytoalexins, very often observed in surviving tissue adjacent to the site of HR, does not cause the killing. This was originally proposed because of the antibiotic effect of phytoalexins only at higher concentrations. Accordingly, so far in no carefully analyzed case could it be shown that the rapid rise in phytoalexin synthesis occurring after pathogen infection is the cause of hypersensitive cell death. Furthermore, after infection of cereals with rust fungi, a typical hypersensitive cell death also occurs, but phytoalexins are not synthesized.The same was found with the bean rust *Uromyces phaseolicola*, which induces HR without accompanying phytoalexin synthesis. Race-specific resistance expressed as HR has also been called an homologous hypersensitive reaction, because it is due to the interaction of a plant with an homologous pathogen.

In most examples an homologous defense reaction causes not only necrosis of a single or a few neighboring cells, but also releases synthesis of several defense related- and pathogenesis related proteins (PR-proteins; see Chap. 5) in metabolically active but more distantly located cells. In addition to the enzymes for phytoalexin synthesis, genes engaged in phenol metabolism and the synthesis of peroxidases, glucanases, chitinases and hydroxyproline-rich glucoproteins are activated. Different synthetic activities are turned on in temporal succession and with increasing distance from the infection site. This suggests that, in the course of an hypersensitive reaction, several different cascades of signal substances are produced, releasing synthetic activities in cells immediately bordering the infection and in some that are more distantly located. Hence, an HR has to be understood as a response to pathogen attack by both the directly afflicted single plant cell and the tissue into which the emerging signals spread. The hypersensitive responses that spread into more distant tissue no longer exhibit programmed cell death, the main characteristic of HR. Rather, they are primed to express plant defense genes. The occurrence and spread of the HR after infection can be monitored by observing the fluorescence of newly synthesised phenolic compounds under UV.

Hypersensitive defense reactions may also occur in some non-host plants after attack by non-homologous pathogens. Such cases of heterologous HR are relatively rare and may be released by parasites equipped with pathogenicity genes which do "not fit" the non-host, perhaps because they do not belong to the appropriate *forma specialis*. Because of the definition of a fungal *forma specialis*, an heterologous HR is classified as an expression of basic resistance. For example, the tobacco plant *Nicotiniana tabacum* can express heterologous HR when challenged with heterologous phytopathogenic bacteria or *Phytophthora infestans*. (Heterologous HR is sometimes also called heterologous incompatibility. However, this is inconsistent defined term since the designations compatibility and incompatibility are reserved for interactions of host plants with their homologous pathogens.) In contrast to homologous HR, the expression of heterologous HR is not tied to the expression of parasite pathogenicity genes. This permits the study of plant stress reactions triggered by an infecting pathogen in the absence of other reactions that might hide them.

So far only a few experiments have been reported analyzing the genetic control of heterologous HR. This issue is a matter for some concern, since pathogen defense by means of heterologous HR may well determine the host range of a pathogen, i.e., its host species specificity and its *forma specialis* (Heath 1991). In one very carefully analyzed system, gene-for-gene interactions were shown to trigger defense reactions in non-host plants (Tosa 1996). In *Erysiphe graminis* f.sp. *agropyri*, four different avirulence genes, *Ppm10, Ppm11, Ppm14, Ppm15*, were identified by four different corresponding resistance genes in wheat, *Pm10, Pm11*, and *Pm15*, triggering HR and pathogen defense. *E. graminis* f.sp. *agropyri* is restricted by this gene-for-gene recognition to a host range which *excludes* any wheat carrying any of these four genes. All wheat cultivars tested by this investigator were not colonized by *E. graminis* f.sp.*agropyri* not because the pathogen lacks the appropriate pathogenicity genes, but because a gene-for-gene interaction superimposed on existing basic compatibility triggers a hypersensitive defense. Tosa called such a pathogen an "inappropriate *forma specialis*". By contrast, the basic resistance observed, for example, with pea plants against powdery mildew *Erysiphe graminis* f.sp. *tritici* has a much more complex genetic background.

Race-specific resistance, with its typical gene-for-gene response triggering pathogen defense, has been observed so far mainly for interactions between plants and biotrophic pathogens such as *Puccinia graminis*, stem rust, on cereals, or hemibiotrophic pathogens such as *Phytophthora infestans*, late blight, on potato. However, very rarely, gene-for-gene interactions eliciting race-specific resistance have also been observed with necrotrophic pathogens, for example, *Rhynchosporium secalis*, scald, on rye and barley. In the latter case, expression of race-specific resistance depends on a very uncommon behavior of a necrotrophic pathogen, namely that, after infection, *R. secalis* does not instantly kill its host plant. Thereby the plant is given a chance to express newly activated defense reactions against the invading pathogen.

Gene-for-gene interactions for triggering defense reactions were also postulated for plant pathogenic viruses, bacteria and insects. In several experimentally examined cases, triggering of gene-for-gene defense reactions seemed very likely, and in a few cases gene-for-gene interactions have even been proved. In biology, gene-for-gene interactions are by no means unusual. One classic and instructive example is the specific adsorption of bacteriophage T2 to its host bacterium *Escherichia coli* to safeguard infection and injection of its DNA. The requirements for this specific interaction between both components are genetically determined by a single gene in each partner. A mutation in either gene prevents interaction.

It is not easy to prove the occurrence of a strict gene-for-gene interaction in the release of race-specific resistance in all its complexity. Besides identifying the individual *Avr* and *R* genes by crossing experiments, identification, isolation, and characterization of the corresponding gene products are also required. This goal can be reached only with sufficient knowledge of the genetics of both partners, the conditions for correct use of molecular genetic methods, and an experimental system for genetic transformation, optimally for both partners.

Such conditions were recently met in two systems, proving the validity of a strict gene-for-gene relationship for releasing race-specific resistance. Single components of the signal-sensor system of the biotrophic pathogen *Cladosporium fulvum* on tomato and the necrotrophic pathogen *Rhynchosporium secalis* on barley were identified and analyzed using molecular genetic methods. Fortunately, the signal substance in both systems was stable, of low molecular weight and diffusable and could easily be isolated and separated from macromolecular cell components. Indeed, the gene-for-gene hypothesis by no means demands that the signal substance is diffusable or that the interaction between signal, or elicitor, and sensor, or receptor, is necessarily direct. Fortunately, in both cases the elicitors were polypeptides and hence primary gene products.

In the system studied by de Wit and his collaborators, the tomato pathogen, *Cladosporium fulvum* (synonym: *Fulvia fulvum*), grows within the intercellular spaces of the leaves, between the mesophyll cells. Conidiophores emerge from the stomata producing large numbers of spores. The conidiophores form large patches on the undersides of the leaves which appear as though the fungus is growing on the surface. However, in the mesophyll the pathogen does not form haustoria as organelles for nutrition. Hence, the pathogen's hyphae gain their nutrients out of the plant cells by some as yet unidentified process, presumably there is a pathogen-induced change in plant cell permeability. Proteins were isolated from intercellular washing fluids of infected leaves (de Wit et al. 1986, 1989) which were synthesized by the pathogen either constitutively or after induction by the plant. The proteins and polypeptides, constitutively formed by the pathogen proved to be factors which are most likely involved in establishing basic compatibility. Hence they were classified as pathogenicity factors (ECP1 and ECP2, see Chap. 7; van den Ackerveken et al. 1993, de Wit et al. 1994).

Another polypeptide is apparently formed only after induction *in planta* and only by a pathogen race avirulent on tomato cultivars harboring a gene determining race-specific resistance. The defense reaction expressed by the plant is an hypersensitive reaction and programmed cell death leads to necrotization of the afflicted leaf tissue. The polypeptide was obtained in relatively large quantities from intercellular washing fluids of susceptible cultivars, e.g., of genotype *Cf0* infected with pathogen races avirulent on tomato cultivars carrying resistance gene *Cf9*. The isolated polypeptide, after infiltration into leaves, causes an HR only in *Cf9* race-specific resistant cultivars. Cultivars harboring resistance genes *Cf2*, *Cf4* and *Cf5* are not affected. Hence, triggering the necrotic resistance reaction required a gene-for-gene interaction, *Cf9* resistance gene in the plant and *Avr9* gene in the pathogen. Furthermore it was suggested that the polypeptide synthesized after induction might be the product of the pathogen's avirulence gene *Avr9* (van Kan et al. 1991, van den Ackerveken et al. 1992). This proposed single avirulence gene could not be identified by classic genetic crossing procedures, because *C. fulvum* has no sexual stage. However, the gene was identified indirectly and cloned. The polypeptide supposedly encoded by gene *Avr9* and functioning as a race-specific elicitor contained 28 amino acids, as revealed by sequencing experiments. Based upon this result, a corresponding oligonucleotide probe was synthesized and a hybridizing cDNA clone selected from a gene expression library obtained from *C. fulvum*. With the cDNA clone homologous to the presumed *Avr9* gene, a second DNA clone was selected by DNA hybridization from a *C. fulvum* genomic library. However, the selected gene encoded a much longer polypeptide than the isolated elicitor, namely 63 amino acids, and its coding sequence contained one intron of 59 base pairs. After removal of its signal peptide, the primary translation product of 63 amino acids is processed to a length of 28 amino acids (van den Ackerveken et al. 1993). This polypeptide, which most probably represents the race-specific elicitor, was recently shown to bind to plasma membranes of several plants in the Solanaceae (Kooman-Gersmann et al. 1996).

The race-specific resistance gene of the tomato should encode the hypothetical corresponding receptor. The *Cf9* gene was very recently cloned by the group of J.D.G. Jones (see Chap. 9, Sect. 9.1.4), so that both partners of the gene-for-gene system triggering race-specific resistance have now been isolated. This provides the opportunity to analyze the plant components participating in recognition.

Some naturally occurring pathogen races of *C. fulvum* are specifically virulent on *Cf9*-resistant cultivars. These races do not owe their virulent behavior to a mutation converting the gene *Avr9* to the defective allele *avr9*. Rather, the virulent behavior is caused by loss of the whole *Avr9* gene. How this loss may have happened is so far unknown. One possibility is that the *Avr9* gene is located on a plasmid which was lost. (Plasmids are extrachromosomal elements which may segregate out spontaneously without their loss being lethal to the cell as they carry no essential genetic information.) Such naturally occurring specific virulent races of *C. fulvum* became avirulent on *Cf9*-resistant tomato cultivars if they carried copies of the *Avr9*-gene introduced by genetic transformation effecting integration of the copies at different chromosomal loci (Van den Ackerveken et al. 1992).– The expression of race-specific resistance in a *Cf9*-resistant tomato cultivar can also be prevented by substituting the *pyrG* gene of *Aspergillus nidulans* for the gene *Avr9* in the pathogen by employing the technique of gene disruption (Marmeisse et al. 1993). This prevents the pathogen from synthesizing the Avr9 elicitor which would trigger defense reactions in the plant.– Finally, it was shown that different single point mutations in another avirulence gene of *C. fulvum*, *Avr4*, corresponding to resistance gene *Cf4* of tomato, also caused virulence of the pathogen mutant: At each of three different positions, a single base exchange leads to incorporation into the *Avr4* gene product of the amino acid tyrosine instead of cysteine, inactivating the elicitor polypeptide.

Several other observations are of interest: (1) In vitro, *C.fulvum* synthesizes the *Avr9* product only during periods of nitrogen starvation. The addition of plant extracts to the culture medium does not result in elicitor synthesis. Also *in planta* elicitor synthesis is very likely induced by nitrogen starvation, since in front of the *Avr* coding region and its promoter are six adjacent nucleotide motifs of the same type found in *Neurospora crassa*. This motif binds the regulator protein NIT2 for positive regulation of nitrogen metabolism (Van den Ackerveken et al. 1994). (2) The plant contributes to production of the elicitor as isolated from the intercellular washing fluid, since a protease produced by the plant processes the 63 amino acid translation product through intermediary peptides of 34, 33 and 32 amino acids to the end-product of 28 amino acids. (3) Experiments employing GUS-reporter gene constructs for monitoring gene expression showed that the *Avr9* gene is expressed immediately after the pathogen hyphae enter the leaf interior through stomata. (4) The three-dimensional structure of the *Avr9* protein is necessary for its elicitor activity: Removal of disulfide bridges generated by six cysteine residues present in the polypeptide destroys elicitor activity, whereas other amino acid exchanges can either decrease or increase activity (Kooman-Gersmann et al. 1997).

Similar results were obtained by de Wit's group analyzing another peptide secreted intercellularly, the product of the *Avr4* avirulence gene of *C. fulvum* (Joosten et al. 1994). Althgough the mechanism of induction of *Avr4* expression *in planta* is still unknown, the primary gene product of 135 amino acids is also processed, via an intermediary product of 117 amino acids, to the final elicitor peptide. Each of seven spontaneous, specific virulent mutants of *Avr4* harbored a single point mutation within the coding region. Each mutation was located at one of three different positions in the *Avr4* coding region, effecting an amino acid exchange in the putative translation product from cysteine (codon TGT) to tyrosine (TAT). Southern- and northern blots with wild type DNA showed that the mutant DNA is transcribed into mRNA. Since cysteine residues are often involved in the formation of disulfide bridges and hence establish tertiary protein structure, the avirulence gene product seems to require a particular three-dimensional structure that is recognized by the plant receptor. Other mutants had an amino acid exchange from tyrosine to histidine, from threonine to isoleucine, or a reading frame shift. The resulting specific virulence in all the mutants was accompanied by several distinct morphologic phenotypes. These were most probably caused by differences in sensitivity of the mutant product to plant proteases and by the lack of recognition at the plant receptor. In particular it was shown that all of these point mutations in *Avr4* make the elicitor unstable, abolishing its activity when infecting *Cf4*-resistant *C. fulvum* mutants. However, necrotic activity may still be released by the defective elicitors when the mutant *avr4* genes are introduced into the plant cell by PVX virus constructs PVX::*avr4* which synthesize its product inside the plant cell (Joosten et al. 1997). It is assumed that under these conditions the mutant elicitors are less exposed to attack by proteases present in fungal and plant cell walls.

The tomato plant infected with *C. fulvum* is the first host-parasite system to demonstrate unequivocally that: (1) a single base exchange can cause specific virulence of a pathogen, and (2) an isolated avirulence factor produced by one particular avirulence gene induces HR on plant leaves. By contrast, the primary gene products of the avirulence genes of phytopathogenic bacteria cloned so far do not trigger HR in the corresponding race-specific resistant host plants, indicating that with these bacteria the connections to elicitor action seem to be more indirect.

In the second system, barley infected by the necrotrophic pathogen, *Rhynchosporium secalis*, Knogge et al. analyzed supposed gene-for-gene interactions. A resistant barley cultivar was isolated expressing monogenically determined resistance against a particular race of *R. secalis* (Knogge et al. 1991; Hahn et al. 1994). Thus this pathogen race exhibits avirulent behavior against one particular resistant barley cultivar. By contrast, a nearly isogenic cultivar is parasitized by the same pathogen race, producing necrotic lesions indicating virulent behavior. In liquid culture this pathogen race excretes three toxic peptides, NIP1, NIP2, and NIP3, into the medium, which produce necrosis when applied to healthy leaves. NIP1, NIP2, and NIP3 are virulence factors coded by the pathogenicity genes *nip1, nip2, and nip3*. These virulence factors were classified as non-host-selective toxins (see Chap. 8, Sect. 8.1). Gene *nip1* (called *avrRrs1* in the original paper) codes for the toxic peptide NIP1, which apparently can act in the resistant barley cultivar like an avirulence factor releasing pathogen defense. The peptide induces synthesis of several proteins including peroxidases, thaumatins, and pathogenesis related proteins like those observed in active basic resistance. As a rule, avirulent behavior of a biotrophic pathogen infecting a correspondingly resistant host plant is generally observed as an HR. However, in this instance of a necrotrophic pathogen, which after infection *does not* instantly kill its host cells (see also Chap. 9, Sect. 9.1) the excreted NIP1 peptide is apparently recognized by the *Rrs1* gene product of the resistant barley cultivar, releasing defense reactions after a short period of subcuticular growth of the mycelium. Under these circumstances the toxic NIP1 product functions like an elicitor of race-specific resistance, and there seems to be a gene-for-gene relationship for recognition between the product of gene *nip1* (or formally gene *AvrRrs1*) in the pathogen and the product of gene *Rrs1* in the host plant.

The role of the NIP1 peptide as a non-host-selective toxin seems to be mainly one of slowly killing the plant cells. At the beginning of the infection process the pathogen colonizes susceptible and resistant plants equally. However, some time later, vigorous mycelial growth between the outer cell wall and the plasmalemma of the epidermis, occurs only in susceptible plants. This leads to an expansive subcuticular stroma with formation of spores causing the appearance of necrotic lesions after 10–14 days. The growth is apparently rendered possible by the toxic NIP1 product, which acts like a virulence factor effecting the release of metabolites from the slowly poisoned plant cells (Rohe et al. 1995). By contrast, in resistant plants, growth of the pathogen mycelium is arrested after some time because the toxic peptide is recognized as an elicitor by the product of gene *Rrs1* which releases the expression of defense genes.

The *nip1* gene exists in only one copy and consists of two exons and one intron of 65 bp. The gene codes for a product of 82 amino acids which is processed by removal of a signal sequence of 22 amino acids. The mature 60-amino acid protein not only acts as a non-host-specific toxin but also as elicitor if a corresponding race-specific resistance gene is present in the plant. The protein contains ten cysteine residues, much like the elicitors of the *Avr9* and *Avr4* genes, which contain six and eight cysteine residues, respectively. Apparently, cysteine residues maintain the three-dimensional structure of these proteins that function as elicitors. One mutant pathogen, specifically virulent against *Rrs1* resistant barley cultivars was isolated in the field. This mutant had eight nucleotide changes in its *nip1* gene however, none of them resulted in loss of a cysteine residue. The mutant NIP1 peptide produced showed the same virulence as the wild type non-host-selective toxin NIP1 synthesized by the avirulent wild type pathogen. The same holds for other naturally occurring mutants with other nucleotide exchanges in the *nip1* gene, effecting either partial or complete loss of elicitor activity for releasing defense reactions in the *Rrs1*-resistant barley cultivar. Apparently, the domains in the NIP1 product that determine elicitor specificity and the domains that determine non-host-specific-toxin activity are located in different regions of the molecule. The wild type NIP1 protein represents the only case known so far in which a virulence factor and a non-host-specific toxin, respectively, can also serve, in the presence of a corresponding resistance gene, as an elicitor for race-specific resistance (compare this with the definitions for virulence and specific virulence in Chap. 1 and Chap. 9, Sect. 9.1.1).

The function of the *Rrs1* gene product has not yet been identified. It could function as a sensor or receptor for the NIP1 peptide, but it could also link the specific recognition reaction to the battery of defense reactions of active basic resistance. However, no details are yet known about the nature of the plant's recognition reaction or the nature of the sensor.

The different elements proposed for the signal-sensor system in *Cladosporium fulvum* on tomato have been convincingly demonstrated at the molecular genetic level, largely because of the progress made in cloning the participating race-specific resistance genes (see Chap. 9, Sect. 9.1.4). Nevertheless, we should not ignore the fact that we are just at the beginning of unraveling the genetic structure and function of race-specific resistance genes to learn more about their biochemistry and molecular biology. Many questions remain about the receptor protein, as proposed in the original hypothesis of an elicitor-receptor reaction, and how best to design future experiments. Moreover, there are arguments against a literal acceptance of the elicitor-receptor hypothesis. These objections have led to other models to explain formal gene-for-gene interactions (see Chap. 9, Sects. 9.1.3.2, 9.1.3.3, 9.1.3.4 and 9.1.4).

Arguments against the elicitor-receptor model to explain the release of race-specific resistance stem from some weaknesses and inconsistencies within the hypothesis and the fact that it does not provide answers to several obvious questions. The most prominent arguments against the elicitor-receptor model are as follows:

(1) Hypersensitive defense reactions (HR) are observed not only with race-specific resistant host plants defending biotrophic pathogens, but also with non-host plants expressing heterologous HR following pathogen attack. Heterologous HR contradicts a basic assumption of the elicitor-receptor model that the induced defense mechanisms are secondary in nature and superimposed on a basic compatibility which has to be established previously. However, interpretation of heterologous HR as a secondary defense mechanism would not make much sense, because non-host plants already express basic resistance. Hence releasing additional defense mechanisms is unnecessary. Moreover, investigations on obligate biotrophic pathogens (e.g., *Erysiphe graminis* f.sp. *agropyri* and wheat, see above results of Tosa 1996) suggest that heterologous HR is qualitatively not different from homologous HR released by race-specific resistance, and that in both cases the same resistance genes and receptors may be involved. Thus it is quite possible that heterologous HR is triggered, at least in some cases, also by a gene-for-gene interaction. – The hypothesis that elicitor-receptor reactions releasing race-specific resistance are necessarily bound to existing basic compatibility leads to another implausible consequence: The pathogen would induce, when triggering the expression of race-specific resistance, the defense mechanisms of active basic resistance which it had already broken down or was able to negate. Why wouldn't the pathogen overcome these secondarily established defense mechanisms again, just as it previously overcame the same mechanisms present at the level of basic resistance?

(2) More than a dozen avirulence genes of phytopathogenic bacteria and fungi have been cloned and characterized so far. However, only in three cases could the synthesis of an elicitor be demonstrated, either directly or indirectly, to be based on a corresponding DNA coding sequence. In all other cases the function of the DNA sequences determining avirulence remains unknown. Three well-known avirulence genes synthesizing a product are: (a) the avirulence gene *avrD* of *Pseudomonas syringae* pv. *glycinea*, (b) the genes discussed above, *Avr9* and *Avr4* of *Cladosporium fulvum*, each forming one peptide, and (c) the gene *AvrRrs1* from *Rhynchosporium secalis*, also discussed above. The *Avr* products or signal substances noted in (b) and (c) were isolated from sterile cultural filtrates and trigger, as specific elicitors in plant leaves, the formation of necrotic lesions. For all other *Avr* genes,

even those with known DNA sequences, no direct connection to an elicitor substance has been proven so far.

(3) No receptor, or receptor protein, able to recognize an avirulence factor has yet been isolated. However, genes coding for race-specific resistance have recently been cloned and sequenced. Sequence comparisons show that several domains can be identified at the DNA level (see also Chap. 9, Sect. 9.1.4) that could accommodate particular structural-functional features. These results make clear that the respective receptor proteins appear to be much more complex than the original elicitor-receptor hypothesis proposed.

(4) The elicitor-receptor model, in its original formulation, does not explain how recognition between elicitor and receptor effects gene expression.

Reviews

Aist J.R., Bushnell W.R. (1991): Invasion of plants by powdery mildew fungi, and cellular mechanisms of resistance. In: Cole G.T., Hoch H.C. (eds.): The Fungal Spore and Disease Initiation in Plants and Animals. Plenum Press, New York, London. 321–345

Atkinson M.M. (1993): Molecular mechanisms of pathogen recognition by plants. Advances in Plant Pathology 10: 35–64

Bailey J.A. (1983): Biological perspectives of host-pathogen interactions. In: Bailey J.A., Deverall B.J. (eds.): The Dynamics of Host Defence. Academic Press Australia, Sidney, New York, London, Paris, San Diego, San Francisco, Sao Paulo, Tokyo, Toronto. 1–32

Boller T. (1995): Chemoreception of microbial signals in plant cells. Annu.Rev.Plant Physiol.Plant Mol.Biol. 46: 189–214

Collmer A., Keen N.T. (1986): The role of pectic enzymes in plant pathogenesis. Annu. Rev.Phytopathol. 24: 383–409

Darvill A.G., Albersheim P. (1984): Phytoalexins and their elicitors – a defense against microbial infection in plants. Annu.Rev.Plant Physiol. 35: 243–275

De Wit P.J.G.M., Joosten M.A.H.J., Honée G., Wubben J.P., van den Ackerveken G.F.J.M., van den Broek H.W.J. (1994): Molecular communication between host plant and the fungal tomato pathogen Cladosporium fulvum. Antonie van Leeuwenhoek 65: 257–262

De Wit P.J.G.M. (1992): Molecular characterization of gene-for-gene systems in plant-fungus interactions and the application of avirulence genes in control of plant pathogens. Annu.Rev.Phytopathol. 30: 391–418

De Wit P.J.G.M., Van den Ackerveken, Vossen P.M.J., Joosten M.H.A.J., Cozijnsen T.N., Honée G., Wubben J.P., Danhash N., Van Kan J.A.L., Marmeisse R., Van den Broek H.W.J. (1993): Avirulence genes of the tomato pathogen Cladosporium fulvum and their exploitation in molecular breeding for disease resistant plants. In: Fritig B., Legrand M. (eds.): Mechanisms of Plant Resistance Responses. Kluwer Academic Publishers, The Netherlands. 24–32

De Wit P.J.G.M., Van Kan J.A.L., Van den Ackerveken A.F.J.M., Joosten M.H.A.J. (1991): Specificity of plant-fungus interactions: Molecular aspects of avirulence genes. In: Hennecke H., Verma D.P.S. (eds.): Advances in Molecular Genetics of Plant-Microbe Interactions. Kluwer Academic Publishers, Dordrecht, The Netherlands. 233–241

Dixon R.A., Lamb C.J. (1990): Molecular communication in interactions between plants and microbial pathogens. Annu.Rev.Plant Physiol. 41: 339–367

Ebel J., Cosio E.G. (1994): Elicitors of plant defense responses. Int.Rev.Cytology 148: 1–36

Ebel J., Grisebach H. (1988): Defense strategies of soybean against the fungus Phytophthora megasperma f.sp. glycinea: a molecular analysis. Trends Biochem.Sci. 13: 23–27

Gabriel D.W. (1989): Genetics of plant parasite populations and host-parasite specificity. In: Kosugue T., Nester E.W. (eds.): Plant-Microbe Interactions: Molecular and Genetic Perspectives. Macmillan, New York. 343–379

Gabriel D.W., Rolfe B.G. (1990): Working models of specific recognition in plant-microbe interactions. Annu.Rev.Phytopathol. **28**: 365 – 391

Goodman R.N., Novacky A.J. (1994): The Hypersensitive Reaction in Plants to Pathogens, A Resistance Phenomenon. APS Press, St.Paul, Minnesota

Heath M.C. (1991): Evolution of resistance to fungal parasitism in natural ecosystems. New Phytol. **119**: 331 – 343

Keen N.T. (1982): Specific recognition in gene-for-gene host-parasite systems. Advances in Plant Pathology **1**: 35 – 81

Keen N.T. (1990): Gene-for-gene complementarity in plant-pathogen interactions. Annual Review of Genetics **24**: 447 – 463

Keen N.T. (1992): The molecular biology of disease resistance. Plant Mol.Biol. **19**: 109 – 122

Keen N.T. (1993): An overview of active disease defense in plants. In: Fritig B., Legrand M. (eds.): Mechanisms of Plant Defense Responses. Kluwer Academic Publishers, The Netherlands. 3 – 11

Knogge W., Marie C. (1997): Molecular characterization of fungal avirulence. In: Crute I.R., Holub E.B., Burdon J.J. (eds.): The Gene-for-Gene Relationship in Plant-Parasitic Interactions. CAB International, Oxon UK, New York NY. 329 – 346

Moerschbacher B.M., Reisener H.-J. (1997): The hypersensitive resistance reaction. In: Hartleb H., Heitefuss R., Hoppe H.-H. (eds.): Resistance of Crop Plants against Fungi. Gustav Fischer, Jena, Stuttgart, LÅbeck, Ulm. 126 – 158

Ryan C.A., Farmer E.E. (1991): Oligosaccharide signals in plants: a current assessment. Annu.Rev.Plant Physiol. **42**: 651 – 674

Scheel D., Parker J.E. (1990): Elicitor recognition and signal transduction in plant defense gene activation. Z.Naturforsch. **45c**: 569 – 575

Thompson J.N., Burdon J.J. (1992): Gene-for-gene coevolution between plants and parasites. Nature **360**: 121 – 125

Tosa Y. (1996): Gene-for-gene relationships in *forma specialis-genus* specificity of cereal powdery mildews. In: Mills D., Kunoh H., Keen N.T., Mayama S. (eds.): Molecular Aspects of Pathogenicity and Resistance: Requirement for Signal Transduction. The American Phytopathological Society, St. Paul, Minnesota. 47 – 55

Van den Ackerveken G.F.J.M., De Wit P.J.G.M. (1995): The *Cladosporium fulvum*-tomato interaction, a model system for fungus-plant specificity. In: Kohmoto K., Singh U.S., Singh R.P. (eds.): Pathogenesis and Host Specificity an Plant Diseases – Histopathological, Biochemical, Genetic and Molecular Bases. Elsevier Science, Oxford, New York, Tokyo. 145 – 160

Van Kan J.A.L., Joosten M.H.A.J., van den Ackerveken G.F.J.M., de Wit P.J.G.M. (1994): Molecular characterization of avirulence determinants of the tomato pathogen *Cladosporium fulvum*. In: Kohomoto K., Yoder O.C. (eds.): Host Specific Toxin: Biosynthesis, Receptor and Molecular Biology. Faculty of Agriculture, Tottori University, Sogo Printing and Publishing Co., Ltd., Tottori, Japan. 251 – 261

Relevant papers

Davis K.R., Darvill A.G., Albersheim P. (1986): Host-pathogen interactions. XXX. Characterization of elicitors of phytoalexin accumulation in soybean released from soybean cell walls by endopolygalacturonic acid lyase. Z.Naturforsch.C **41c**: 39 – 48

Davis R.D., Lyon G.D., Darvill A.D., Albersheim P. (1984): Host-pathogen interactions XXV. Endopolygalacturonic acid lyase from *Erwinia carotovora* elicits phytoalexin accumulation by releasing plant cell wall fragments. Plant Physiol. **74**: 52 – 60

De Wit P.J.G.M., Buurlage M.B., Hammond K.E. (1986): The occurence of host-, pathogen- and interaction-specific proteins in the apoplast of *Cladosporium fulvum* (syn.*Fulvia fulva*) infected tomato leaves. Physiol.Mol.Plant Path. **29**: 159 – 172

De Wit P.J.G.M., van den Ackerveken G.F.J.M., Joosten M.H.A.J., van Kan J.A.L. (1989): Apoplastic proteins involved in communication between tomato and the fungal pathogen *Cladosporium fulvum*. In: B.J.J. Lugtenberg (ed.): Signal Molecules in Plant and Plant-Microbe Interactions. Springer- Verlag, Berlin, Heidelberg. 273 – 280

Fritzemeier K.-H., Cretin C., Kombrink E., Rohwer F., Taylor J., Scheel D., Hahlbrock K. (1987): Transient induction of phenylalanine ammonia-lyase and 4-coumarate:CoA ligase mRNA in potato leaves infected with virulent or avirulent races of *Phytophthora infestans*. Plant Physiol. **85**: 34 – 41

Hahn M., Jüngling S., Knogge W. (1993): Cultivar-specific elicitation of barley defense reactions by phytotoxic peptide NIP1 from *Rhynchosporium secalis*. Molec.Plant-Microbe Interact. **6**: 745 – 754

Heath M.C. (1991): The role of gene-for-gene interactions in the determination of host species specificity. Phytopathology **81**: 127 – 130

Jin D.F., West A.W. (1984): Characteristics of galacturonic acid oligomers as elicitors of casbene synthetase activity in castor bean seedlings. Plant Physiol. **74**: 989 – 992

Joosten M.A.H.J., Cozijnsen T.J., De Wit P.J.G.M. (1994): Host resistance to a fungal tomato pathogen lost by a single base-pair change in an avirulence gene. Nature **367**: 384 – 386

Joosten M.A.H.J., Vogelsang R., Cozijnsen T.J., Verberne M.C., De Wit P.J.G.M. (1997): The biotrophic fungus *Cladosporium fulvum* circumvents *Cf-4*-mediated resistance by producing unstable AVR4 elicitors. Plant Cell **9**: 367 – 379

Knogge W., Hahn M., Lehnackers H., Råpping E., Wevelsiep L. (1991): Fungal signals involved in the specificity of the interaction between barley and *Rhynchosporium secalis*. In: Hennecke H., Verma D.P.S. (eds.): Advances in Molecular Genetics of Plant-Microbe Interactions. Dordrecht, Kluwer Academic Publishers, Netherlands. 250 – 253

Kooman-Gersmann M., Honée G., Bonnema G., De Wit P.J.G.M. (1996): A high-affinity binding site for the AVR9 peptide elicitor of *Cladosporium fulvum* is present on plasma membranes of tomato and other solanaceous plants. Plant J. **8**: 929 – 938

Kooman-Gersmann M., Vogelsang R., Hoogendijk E.C.M., De Wit P.J.G.M. (1997): Assignment of amino acid residues of the AVR9 peptide of *Cladosporium fulvum* that determinate elicitor activity. Molec. Plant-Microbe Interactions **10**: 821 – 829

Marmeisse R., Van den Ackerveken G.F.J.M., Goosen T., de Wit P.J.G.M., Van den Broek H.W.J. (1993): Disruption of the avirulence gene *avr9* in two races of the tomato pathogen *Cladosporium fulvum* causes virulence on tomato genotypes with the complementary resistance gene *Cf9*. Molec.Plant-Microbe Interact. **6**: 412 – 417

Rohe M., Gierlich A., Hermann H., Hahn M., Schmidt B., Rosahl S., Knogge W. (1995): The race-specific elicitor, NIP1, from the barley pathogen, *Rhynchosporium secalis*, determines avirulence on host plants of the *Rrs1* resistance genotype. EMBO J. **14**: 4168 – 4177

Rohwer F., Fritzemeier K.-H., Scheel D., Hahlbrock K. (1987): Biochemical reactions of different tissues of potato (*Solanum tuberosum*) to zoospores or elicitors from *Phytophthora infestans*. Accumulation of sesquiterpenoid phytoalexins. Planta **170**: 556 – 561

Van den Ackerveken G.F.J.M., Dunn R.M., Cozijnsen A.J., Vossen J.P.M.J., Van den Broek H.W.J., De Wit P.J.G.M. (1994): Nitrogen limitation induces expression of the avirulence gene *avr9* in the tomato pathogen *Cladosporium fulvum*. Mol.Gen.Genet. **243**: 277 – 285

Van den Ackerveken G.F.J.M., Van Kan J.A.L., De Wit P.J.G.M. (1992): Molecular analysis of the avirulence gene *avr9* of the fungal tomato pathogen *Cladosporium fulvum* fully supports the gene-for-gene hypothesis. Plant J. **2**: 359 – 366

Van den Ackerveken G.F.J.M., Vossen P., De Wit J.G.M. (1993): The *AVR9* race-specific elicitor of *Cladosporium fulvum* is processed by endogenous and plant proteases. Plant Physiol. **103**: 91-96

Van Kan J.A.L., Van den Ackerveken G.F.J.M., De Wit P.J.G.M. (1991): Cloning and characterization of cDNA of avirulence gene *avr9* and the fungal pathogen *Cladosporium fulvum*, causal agent of tomato leaf mold. Molec.Plant-Microbe Interact. **4**: 52 – 59

9.1.3.2
The Dimer Model

The dimer model, proposed by Ellingboe in 1982, explains the release of race-specific resistance by assuming, like the elicitor-receptor model, that basic compatibility already exists between host plant and pathogen. The hypothesis emerged from the idea that complex processes could possibly best be analyzed by starting from an hypothesis based on genetic determinants their existence has already been demonstrated. The elicitor-receptor model proposed two types of host plant genes engaged in the release of race specific resistance: those coding for recognition of the attacking pathogen and those coding for the defense mechanisms to be expressed. However, strong genetic evidence for the existence of both types of genes was at that time still lacking. Accordingly, the dimer hypothesis was based solely on genetic data demonstrating that a single gene in the host plant and one in the pathogen determine the expression of race-specific resistance. The expression of race-specific resistance was believed to be due to genetic regulatory mechanisms such as those observed in prokaryotes, by which the transcription of genes coding for reaction chains leading to basic compatibility are switched off.

Like the elicitor-receptor model, the dimer model attributes the observed gene-for-gene relationship to a "recognition" which occurs by the formation of a complex, the so-called "dimer", between the avirulence factor, coded by gene *Avr* of the pathogen, and the resistance factor, coded by gene *R* of the host plant. However, this recognition does not take place at the plant cell membrane, as in the elicitor-receptor model. Instead recognition consists of the formation of a regulatory dimer made up of two gene products synthesized by the "signal gene" *Avr* and the "sensor gene" *R* (*Fig. 7*). This dimer acts as a negatively acting regulator, directly blocking the expression of genes leading to the establishment of basic compatibility and hence to parasitism. Thus, the dimer acts at the level of transcription. The products of the avirulence and resistance genes that form the dimer were proposed to consist either of the corresponding mRNAs, the translated proteins of both genes, or one mRNA and one translated protein of each one of the partners. There was also the possibility that the dimer consisted of the two primary gene products bound to particular site(s) on the DNA, or that from the dimer some regulatory active molecule is cleaved off. The term "dimer" for the hypothetical regulatory active complex is unfortunate, since in biochemistry a dimer is defined as a complex comprising two identical molecules. By contrast, the above discussed model defines the "dimer" as a complex formed by specific recognition between quite different molecules.

Once basic compatibility is blocked by the dimer, basic resistance is restored. This is in contrast to the elicitor-receptor model, which assumes that new defense mechanisms are established by the host plant in the presence of basic compatibility. The dimer was also proposed to release hypersensitive cell death (see

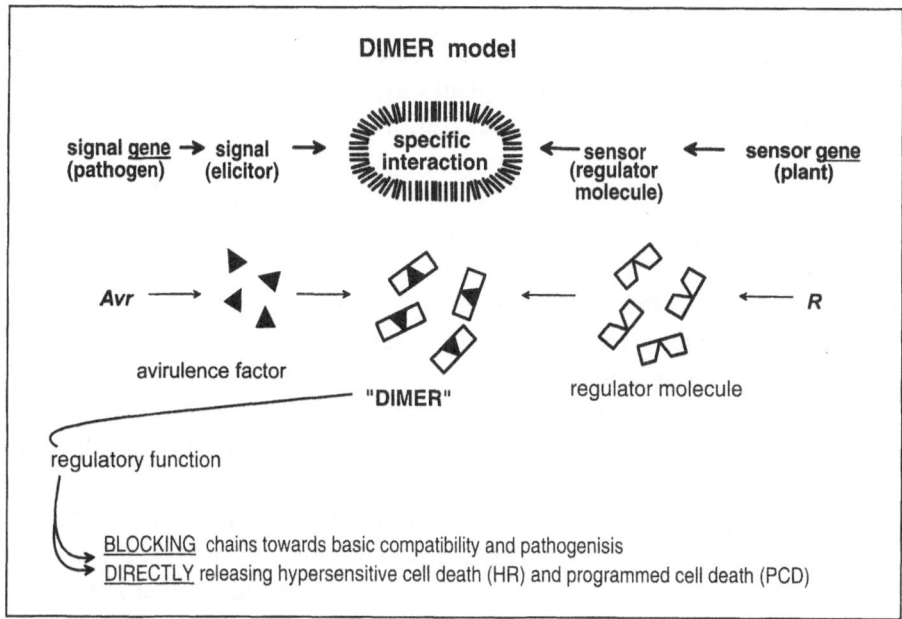

Figure 7. The dimer model for gene-for-gene interaction. The *upper* part of the figure depicts the hypothetical signal-sensor reaction. The *lower* part shows the most essential features of the dimer-model: (1) the avirulence factor as a pathogen-derived signal molecule, or elicitor, produced by the avirulence gene *Avr*, and (2) the plant-derived sensor, or resistance factor, as the regulator molecule coded for by resistance gene *R*. Recognition between the signal and sensor molecules effects the formation of the regulatory active complex, the dimer, effecting basic compatibility by blocking transcription of genes coding for plant defense mechanisms. For details, see text

Fig. 7), but no mechanism for this function was proposed. The restoration of defense mechanisms belonging to basic resistance by the action of the dimer may be called a "reactivated defense".

There are two main differences between the dimer- and elicitor-receptor models: First, the dimer model is based on the idea that the defense mechanisms of basic resistance are "reactivated" by the regulatory action of the dimer, unlike the elicitor-receptor model in which new defense mechanisms are established in the presence of basic compatibility. Second, specific recognition between avirulence and resistance factor gives rise to the dimer, which acts as a genetic regulator directly at the DNA level. In the elicitor-receptor model, after specific recognition at the cell membrane, additional reaction chains release the expression of defense genes in the cell nucleus. In other words, pathogen defense or incompatibility, according the dimer model, is believed to result from a block of reaction chains involved in basic compatibility or pathogenicity, whereas the elicitor-receptor model regards incompatibility as the establishment of new defense mechanisms in the presence of already existing basic compatibility.

The dimer models offers a rather simple explanation for heterologous incompatibility, since the same regulatory events could cause homologous as well as heterologous incompatibility, namely, the blockade of reaction chains leading to basic compatibility. This explanation would avoid difficulties arising from the elicitor-receptor model that try to explain why, after establishment of basic compatibility, the plant again defends itself against attack, this time at a different level but by the same mechanisms the pathogen has just overcome when breaking down basic resistance. Furthermore, according to the dimer model, the different race-specific resistance genes $(R_1, R_2,....R_n)$ or their products could form, together with the corresponding avirulence factors, dimers blocking distinct reaction chains. This could help in understanding the observed diversity of defense reactions, such as the different types of HR, or differences in the prevention of penetration and haustoria formation.

However, the dimer model is also at variance with some observations and leaves essential questions unanswered: (1) In some exceptional cases oligogenic determination of avirulence was observed (see also Chap. 9, Sect. 9.1.8, "gene-for-gene-pathway"), i.e., more than one gene participates in determination of avirulence; these would be difficult to reconcile with the mechanism of the dimer model. (2) None of the avirulence genes so far sequenced exhibits, either in their nucleotide or derived amino acid sequences, recognition structures suitable for the formation of dimers, i.e., complexes consisting of nucleic acid-protein, nucleic acid-nucleic acid, or protein-protein. (3) Formation of a dimer has so far not been demonstrated experimentally. (4) How the dimer actually causes incompatibility remains unclear. – We may summarize by saying that there is little or no experimental support for the dimer model.

Reviews

Ellingboe A.H. (1982): Genetical aspects of active defence. In: Wood R.S.i. (ed.): Active Defence Mechanisms in Plants. Plenum Press, New York. 179–192

Gabriel D.W., Rolfe B.G. (1990): Working models of specific recognition in plant-microbe interactions. Annu.Rev.Phytopathol. **28**: 365–391

9.1.3.3
The Ion Channel Defense Model

The third model proposed to explain gene-for-gene interactions is the ion channel defense model. It shifts the emphasis of the interaction between the elicitor and the resistance gene product from a more or less immediate effect on gene expression to the epigenetic level. This level represents the network of signal transduction processes that regulate metabolic activities and control development by either activating or blocking gene expression, permanently or transiently. The model, first proposed in 1988 by Gabriel and collaborators, assumes that transmembrane proteins, located at the cell surface, some of which function as ion channels, provide all of the steps necessary for triggering race-specific resistance. Thus the same system would provide not only for pathogen recognition, but also for the transduction of the signals releasing the metabolic activities involved in pathogen defense. The model provides a plausible explanation of several phenomena observed in race-specific resistance such as the release of HR, hypersensitive cell death, and the activation of plant defense genes in active basic resistance (see below).

The ion channel defense model was inspired by findings obtained from electrophysiological experiments on plants and other organisms. These experiments showed that electrical membrane potentials and ion concentrations in the cytoplasm are key elements in the complex signal transduction systems that regulate cell activities. In plants, membrane-bound ion pumps play an essential role in maintaining concentration gradients between the cell interior and the exterior. Ca^{2+} ions play a particularly important part in controlling transcription activation. The Ca^{2+} ion concentration outside the cell, in the cell wall, and within vacuoles ranges from micro- to millimolar, whereas in the cytoplasm may be as much as 10^{-5} lower. A small increase in the cytoplasmic Ca^{2+} concentration may, for example, activate transcription and other metabolic processes. Hence, the integrity of the plasmalemma and tonoplast are of vital importance for the cell and its metabolic activities.

An important reason for proposing the ion channel defense model was the finding that plants subject to stress such as mechanical wounding, or infection, lose electrolytes from the affected cells or tissue. This also occurs if the pathogen produces toxins or elicitors that are specifically recognized by the cell. The first response observed is an efflux or leakage of K^+ ions and an influx of H^+ ions. The loss of electrolytes, accompanied by membrane depolarization, was first believed to be a side effect of the stress applied from wounding and/or cell death. However, there is another interpretation. The loss of electrolytes and membrane depolarization may result from a recognition event associated with wounding or pathogen attack, that ends in hypersensitive cell death and/or the release of other wound and defense reactions. All of these reactions could be activated by the appearance of signal substances produced as a result of the loss of electrolytes from injured cells. According to this interpretation hypersensitive cell death does not cause the loss of electrolytes. Instead the recognition event leads to loss of electrolytes followed by hypersensitive cell death and then to the initiation of other defense reactions in adjacent tissue.

If ion channels in cell membranes are involved in regulating cellular metabolism, the release of race-specific resistance might be described as follows. Recognition between the specific elicitor, the *Avr* product of the pathogen, and its corresponding transmembrane protein in the plant acts as a trigger to open transmembrane proteins linked to ion channels (*Fig. 8*). This model follows the gene-for-gene relationship since a single pathogen gene encodes the elicitor, liberated either directly or indirectly by enzyme action, and a single gene of the plant determines the corresponding transmembrane protein. Opening the ion channels

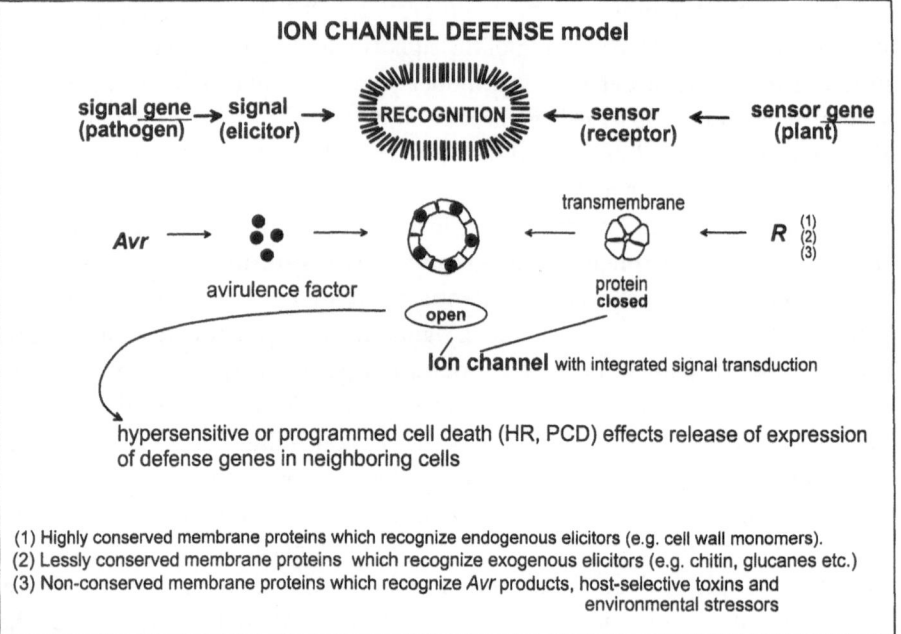

Figure 8. The ion channel defense model for gene-for-gene interaction. The *upper* part of the figure shows the elements involved in the hypothetical signal-sensor reaction. The *lower* part shows the principal elements and processes: (1) the avirulence factor is the signal molecule delivered by the pathogen, and (2) the transmembrane protein is the plant's sensor and ion channel in the cytoplasmic membrane. The interaction between the avirulence factor and the transmembrane protein opens the ion channel, allowing for ion exchange on both sides of the membrane. This exchange alters cellular metabolism leading to the expression of defense reactions. For further details, see text

leads to the efflux of K^+ and Cl^+ ions, to the influx of H^+ and Ca^{2+} ions, the so-called K^+/H^+ response, and to a variable depolarization of the membrane, so that the plasmalemma and tonoplast disintegrate. Complete membrane disintegration would cause instant cell death, but local impairment could well liberate signals which would diffuse into neighboring cells inducing stress reactions in them. Furthermore, the intensity of the signals triggered by recognition between elicitor and transmembrane protein may depend on the size and number of ion channels, the number of available elicitor molecules and their binding intensity to membrane proteins, and the magnitude and speed of substance exchange through the cell membrane.

The basic ion channel defense model was further modified by speculations about the nature and function of the transmembrane proteins. Three kinds of membrane proteins were proposed (*Fig. 8*) that act as receptors of signal substances or elicitors and which function themselves as specific ion channels, or have close functional contact with them. These transmembrane protein receptors are able to recognize different signals and elicitors, and can account for a variety of stress reactions, not only those triggering race-specific resistance. The three types of transmembrane proteins also led to a general hypothesis about the evolution of membrane proteins engaged in plant defense reactions.

The three types of transmembrane proteins can be classified as follows:

1. Membrane proteins coded for by highly conserved plant genes involved in the production of macromolecules such as cellulose. Many proteins and enzymes involved in cellulose biosynthesis are probably able to recognize specific intermediary products formed within this pathway. If incorporated into the cell membrane as transmembrane proteins, they might be able to recognize the same substance outside the cell when it might arise from mechanical wounding of the plant or pathogen attack. In this case the substance recognized by the transmembrane protein would act as an endogenous elicitor, opening corresponding ion channels and releasing the expression of defense reactions.

2. Membrane proteins coded for by less well-conserved plant genes. These serve as receptors that recognize widespread exogenous elicitors originating from pathogens such as chitin, glucans or other polymeric molecules synthesized by highly conserved genes of the pathogen. These polymers are digested to oligomers or monomers by exoenzymes of the plant. Recognition of the resulting digestion products would lead to the opening of corresponding ion channels and release of the expression of defense reactions. Transmembrane or receptor proteins of the first and second class would be involved in the release of active basic resistance.

3. Membrane proteins coded for by nonconserved plant genes. These genes undergo relatively frequent mutation because of duplication and recombination events within the corresponding DNA regions. Each transmembrane protein generated as a result of a mutation would function as a receptor and would be able to recognize a particular molecule formed as a consequence of a variety of environmental stress factors, including molecules active in host plant-pathogen interactions like avirulence factors and host-selective toxins. Again, this would open ion channels. The third class of transmembrane proteins would likely be of recent evolutionary origin (see also Sect. 9.1.4, paragraph (4)). In this way novel receptor or transmembrane proteins could arise, each exhibiting a new specificity for recognizing a particular environmental stressor molecule or an avirulence factor. In the latter case, a new type of race-specific resistance specificity would have been created.

Immediate hypersensitive cell death HR, or programmed cell death (PCD; Sect. 9.1.3.1), results from recognition between an elicitor and its corresponding transmembrane protein. This kind of defense reaction is the fastest and most radical defense mechanism and is generally observed only after attack by biotrophic pathogens (see also Sect. 9.1). In contrast, inducing the synthesis of defense compounds, e.g., phytoalexins, in the surviving tissue surrounding cell(s) expressing PCD is a form of HR that develops more slowly. Defense against biotrophic pathogens is a much more complex process, since offensive and defensive actions compete with each other: the offensive impairment of the plant cell membrane by the attacking pathogen, and the defensive synthesis by the attacked plant of defense-related- and PR-proteins.

As a general rule several different defense reactions are released by a single recognition event. This may be due either to one signal activating different synthetic pathways or to different pathways set in motion by secondary signals produced by the action of the first signal following the recognition event. Most of the synthetic pathways and substances produced are taxon-specific, and found only in particular plant genera or families. Very few are species-specific.

Further experiments confirmed the validity of the ion channel defense model and also modified and deepened our understanding of the trigger mechanisms following recognition that release defense gene expression. Some of the relevant observations came from animal and bacterial cell systems. Applying these findings to phytopathogenic fungi is justified because signal transduction systems are ancient and likely to be highly conserved in evolution. Their mechanisms were probably already present in primitive unicellular eukaryotes before they evolved into the three phyla of fungi, plants and animals.

The most significant modification of the ion channel defense model arose from the finding that the receptor protein and ion channel are not joint structures determined by a single gene. Instead, ion channels are genetically independent protein structures and are not opened by direct interaction with the elicitor. Rather, the elicitor is recognized by an independently encoded receptor protein and the recognition event is then signaled to the ion channels. Hence ion channels are functionally subordinate to the receptor proteins. The latter are the first link in several genetically independent, "downstream" structures activated by signal cascades. These signals consist of the second messengers, cAMP, phosphatidyl inositol, inositol, linoleic acid, lipoxidase metabolites, Ca^{2+}- and H^+ ions, which activate enzymes downstream in the reaction chain. Together with the enzymes, they constitute the signal transduction chain which finally activates the different defense genes.

The signal transduction leading to HR occurs rapidly. If the first signal triggered by the recognition reaction is strong enough, hypersensitive cell death, or instant programmed cell death (PCD), ensues which typically requires protein synthesis. However, if the signal is not strong enough for instant release of PCD, second messengers formed by the penetrated cells may reach neighboring cells either triggering PCD in them, or simply activating the transcription of defense genes. In this way HR mediated by second messengers can spread into the surrounding tissue. A strong signal resulting from recognition also causes production of easily traceable amounts of H_2O_2 or closely related oxidative products (ROI, *reactive oxygen intermediates*, or ROS, *reactive oxygen species*). These compounds, part of the so-called oxidative burst, determine the size of the local PCD-induced necrosis as well as the area of tissue involved in expression of defense genes.

The signal transduction system is thus a complex network of signal chains that interact or even interfere with each other. We are just beginning to understand how the expression of defense genes is controlled (see also Sect. 9.1.4; (2) *Signal transduction*). The following sequence of events appears to be likely: After recognition between elicitor and receptor at the cell membrane, there are changes in ion fluxes through ion channels, e.g., the K^+/H^+ response, uptake of Ca^{2+} ions, and opening of anion channels resulting in a change in membrane potential. In response to these changes, membrane-bound kinases, phosphatases, phospholipases and G proteins are activated. The kinases activate the NADPH oxidase complex, transforming O_2 into ROI such as H_2O_2, O_2^- and HO_2 (Viard et al. 1994). The ROI act directly on the cell wall by incorporating lignin and by cross-linking cell wall proteins. ROI also activate transcription factors in the cytoplasm as well as membrane-bound kinases, phosphatases, phospholipases and G proteins (Vera-Estrella et al. 1994). Subsequently, further signals are generated, that activate PCD, HR and the synthesis of defense-related proteins or enzymes modifying phytoanticipins (see Chap. 6). Specific metabolic pathways are also turned on, producing lipoxides, jasmonic acid, ethylene, additional ROIs, benzoic acid, and especially salicylic acid (Delaney et al. 1994; Lee et al. 1995). These latter products are also involved in the systemic spread of resistance through the plant, the so-called systemic acquired resistance (SAR; see Chap. 11). The sequence of events may become altered somewhat, depending on existing genetic and/or physiological conditions. However, the principle for releasing race-specific resistance remains the same: The recognition of a pathogen at the plant's surface generates a signal which is amplified and transduced, via signal cascades or downstream signaling, to the cell nucleus where genes leading to the expression of defense mechanisms are activated. In *Fig. 9* these events are shown in a simplified schematic form.

Receptors that recognize host-specific toxins probably arose de novo, for example, by spontaneous mutation, as is thought to occur with the appearance of a new receptor specificity of a race-specific resistance gene for recognition of a particular avirulence product. One instructive example of the emergence by mutation of a new receptor specificity is the receptor protein in maize that recognizes toxins of two phytopathogenic fungi, the HMT-toxin of *Cochliobolus heterostrophus*, race T and the PM-toxin of *Phyllosticta maydis*. This may be an example of how new receptor proteins could arise. The mitochondrial DNA fragment TURF 2H3, encoding the receptor protein recognizing both toxins, was shown to have been the result of several recombination events occurring in a particular region of maize mtDNA (Dewey et al. 1986). The resulting reading frame created a gene called *T-urf13*. Its translation product, a 13 kDa protein, is inserted into the mitochondrial membrane where it serves as a receptor for both toxins. The most significant phenotype of the gene *T-urf13* however, is that plants carrying it are male sterile, a feature selected by maize breeders to facilitate the production of F_1 hybrid maize cultivars in agriculture. The sen-

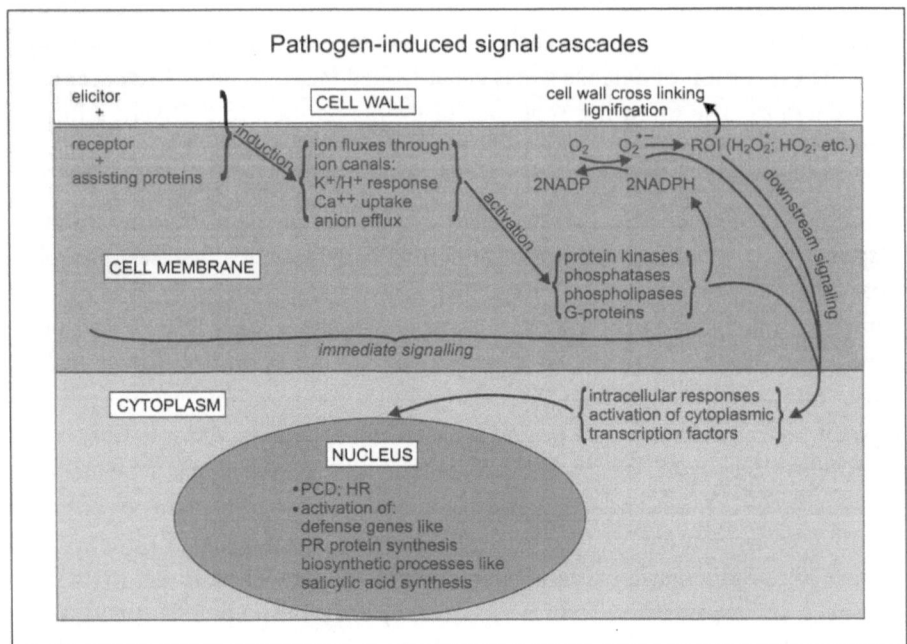

Figure 9. Elicitor-induced signal cascades releasing defense reactions in plant cells. *HR*, hypersensitive reaction; *PCD*, programmed cell death; *ROI*, reactive oxygen intermediates

sitivity of these cultivars to the HMT-toxin was a totally unexpected result, that proved to be devastating for farmers growing hybrids that carried the male sterile cytoplasm when race T of Southern corn leaf blight appeared in 1970.

Similar recombination events might have given rise to the chromosomal locus *Vb/Pc-2* of oat, encoding the receptor recognizing a particular avirulence factor of *Puccinia coronata* (Rines and Luke 1985), and the host-specific toxin victorin of *Helminthosporium victoriae* (see also Chap. 8, Sects. 8.2, 8.3). Victorin also acts in oat as an elicitor for the synthesis of a phytoalexin, avenalumin but, only in the presence of the *Pc-2* allele (Mayama et al. 1986). Whether or not there is a close relationship between the molecular structures of victorin and the avirulence factor of *Puccinia coronata*, and possibly also between the reaction chains released by both molecules in oat plants, is not known.

The ion channel model explains the release of race specific resistance as follows: Class 3 transmembrane or receptor proteins recognize avirulence products as specific elicitors. Recognition provokes the opening of ion channels in the plasma membrane, disturbing its structure and electrical potential. These changes are transduced via signal transduction chains consisting of membrane-bound enzyme complexes and second messengers, thereby activating gene expression both in the cell penetrated by the pathogen and in neighboring tissue. The gene activation may result either in immediate hypersensitive cell death or PCD, and/or expression of genes involved in the erection of new defense barriers, depending on the genetic and/or physiological conditions in the plant.

Reviews

Atkinson M.M. (1993): Molecular mechanisms of pathogen recognition by plants. Advances in Plant Pathology 10: 35–64

Dangl J.L., Dietrich R.A., Richberg M.H. (1996): Death Don't Have No Mercy: Cell Death Programs in Plant-Microbe Interactions. Plant Cell 8: 1793–1804

Dixon R.A., Lamb C.J. (1990): Molecular communication in interactions between plants and microbial pathogens. Annu.Rev.Plant Physiol. 41: 339–367

Ebel J., Grisebach H. (1988): Defense strategies of soybean against the fungus *Phytophthora megasperma* f.sp. *glycinea*: a molecular analysis. Trends Biochem.Sci. 13: 23–27

Gabriel D.W., Loschke D.C., Rolfe B.G. (1988): Gene-for-Gene recognition: The ion channel defense model. In: Verma D.P., Palacios R. (eds.): Molecular Plant-Microbe Interactions. American Phytopathological Society Press, St. Paul. 3–14

Gabriel D.W., Rolfe B.G. (1990): Working models of specific recognition in plant-microbe interactions. Annu.Rev.Phytopathol. 28: 365–391

Hammond-Kosack K.E., Jones J.D.G. (1996): Resistance Gene-Dependent Plant Defense Responses. Plant Cell 8: 1773–1791

Knogge W. (1997): Elicitors and suppressors of the resistance response. In: Hartleb H., Heitefuss R., Hoppe H.-H. (eds.): Resistance of Crop Plants against Fungi. Gustav Fischer, Jena, Stuttgart, Lübeck, Ulm. 159–182

Knogge W. (1998): Fungal pathogenicity. Curr. Opinion Plant Biol. 1: 324–328

Kombrink E., Somssich I.E. (1995): Defense response of plants to pathogens. Adv.Bot.Res. 21: 1–34

Kronstad J.W. (1997): Virulence and cAMP in smuts, blasts and blights. Trends Plant Science 1: 193–199

Lee H.I., Leon J., Raskin I. (1995): Biosynthesis and metabolism of salicylic acid. Proc. Natl.Acad.Sci.USA 92: 4076–4079

Novacky A. (1991): The plant membrane and its response to disease. In: Cole G.T., Hoch H.C. (eds.): The Fungal Spore and Disease Initiation in Plants and Animals. Plenum Press, New York, London. 363–378

Ryan C.A., Farmer E.E. (1991): Oligosaccharide signals in plants: a current assessment. Annu.Rev.Plant Physiol. 42: 651–674

Scheel D., Colling C., Hedrich R., Kawalleck P., Parker J.E., Sacks W.R., Somssich I.E., Hahlbrock K. (1991): Signals in plant defense gene activation. In: Hennecke H., Verma D.P.S. (eds.): Advances in Molecular Genetics of Plant-Microbe Interactions. Dordrecht, Kluwer Academic Publishers, Netherlands. 373–380

Scheel D., Parker J.E. (1990): Elicitor recognition and signal transduction in plant defense gene activation. Z.Naturforsch. 45c: 569–575

Tenhaken R., Levine A., Brisson L.F., Dixon R.A., Lamb C. (1995): Function of the oxidative burst in hypersensitive disease resistance. Proc.Natl.Acad.Sci.USA 92: 4158–4163

Wubben J.P., Boller T., Honée G., De Wit P.J.G.M. (1997): Phytoalexins. In: Hartleb H., Heitefuss R., Hoppe H.-H. (eds.): Resistance of Crop Plants against Fungi. Gustav Fischer, Jena, Stuttgart, Lübeck, Ulm. 202–237

Relevant papers

Delaney T.P., Uknes S., Vernooij B., Friedrich L., Weymann K., Negrotto D., Gaffney T., Gutrella M., Kessmann H., Ward E., Ryals J. (1994): A central role of salicylic acid in plant disease resistance. Science **266**: 1247–1250

Dewey R.E., Levings III C.S., Timothy D.H. (1986): Novel recombinations in the maize mitochondrial genome produce an unique transcriptional unit in the texas male-sterile cytoplasm. Cell **44**: 439–449

Mayama S., Tani T., Uneo T., Midland S.L., Sims J.J., Keen N.T. (1986): The purification of victorin and its phytoalexin elicitor activity in oat leaves. Physiol.Mol.Plant Path. **29**: 1–18

Rines H., Luke H.H. (1985): Selection and regeneration of toxin-insensitive plants from tissue cultures of oats *Avena sativa* susceptible to *Helminthosporium victoriae*. Theor. Appl.Genet. **71**: 16–21

Vera-Estrella R., Barkla B.J., Higgins V.J., Blumwald E. (1994): Plant Defense Response to Fungal Pathogens – Activation of Host-Plasma Membrane H^+-ATPase by Elicitor-Induced Enzyme Dephosphorylation. Plant Physiol. **104**: 209–215

Viard M.P., Martin F., Pugin A., Ricci P., Blein J.P. (1994): Protein Phosphorylation Is Induced in Tobacco Cells by the Elicitor Cryptogein. Plant Physiol. **104**: 1245–1249

9.1.3.4
The Suppressor-Receptor Model

The suppressor-receptor model for the release of race-specific resistance is based on the suppressor model of basic resistance proposed for the establishment of basic compatibility (Chap. 6). Its salient features are that all plants are susceptible to attack by any pathogen, hence all plants exhibit basic compatibility. However, basic compatibility is counteracted by a general elicitor produced by all pathogens which releases unspecific basic resistance. In order to colonize a particular plant, the homologous pathogen has to produce a specific suppressor to block the action of the general elicitor, i.e., the pathogen blocks secondarily its own elicitor of basic resistance. This model was extended by Bushnell and Rowell (1981), and Heath (1982), to race-specific resistance, to account for its gene-for-gene nature, dominant expression, and the observed coevolution of race-specific resistant host plants and specific virulent pathogens. The model was called the suppressor-receptor hypothesis for releasing race-specific resistance.

The model postulated, as its first step, the emergence in the pathogen of a dominantly expressed mutation which modifies the suppressor of the general elicitor but does not disturb the specific binding between the elicitor and its suppressor (or any reactions following this binding). Therefore, this mutation has no detectable effect on the pathogen phenotype and does not interfere with the establishment of basic compatibility. However, as the second step, this mutation may cause a change in the plant's phenotype if the plant carries a dominant mutation in the proposed receptor gene recognizing the pathogen's general elicitor. This corresponds to the appearance of an *R* gene for race-specific resistance. In this case, the mutation in the suppressor would block its interaction with the receptor and prevent interference with the reactions following it, thus effecting incompatibility between pathogen and plant, or pathogen defense. The mutated suppressor that releases race-specific resistance is comparable to the *Avr* product of the elicitor-receptor model. Together, both dominant mutations, one in the pathogen affecting the structure of its specific suppressor and the other in the host plant's receptor blocking its interaction with the modified suppressor (or any further downstream reaction), effect incompatibility and render possible the coevolution of race-specific host plants and specific virulent *avr* pathogens, as has been observed in farming following the large scale introduction of race-specific resistant cultivars (see also Sect. 9.1.5). However, it must be admitted that there is so far no experimental proof for the rather sophisticated appearing suppressor-receptor model.

Reviews

Bailey J.A. (1983): Biological perspectives of host-pathogen interactions. In: Bailey J.A., Deverall B.J. (eds.): The Dynamics of Host Defence. Academic Press Australia, Sidney, New York, London, Paris, San Diego, San Francisco, Sao Paulo, Tokyo, Toronto. 1–32

Heath M.C. (1982): The absence of active defence mechanisms in compatible host-pathogen interactions. In: Wood R.K.S. (ed.): Active Defence Mechanisms in Plants. Plenum Press, New York, London. 143–156

Keen N.T. (1993): An overview of active disease defense in plants. In: Fritig B., Legrand M. (eds.): Mechanisms of Plant Defense Responses. Kluwer Academic Publishers, The Netherlands. 3–11

Relevant papers

Bushnell W.R., Rowell J.B. (1981): Suppressors of defense reactions: A model for roles in specificity. Phytopathology 71: 1012–1014

9.1.4
Plant Genes Conditioning Receptor Function

Both the elicitor-receptor and the ion channel defense models propose that specific membrane-bound proteins, the products of race-specific resistance genes, are receptors that recognize specific pathogen elicitors and that recognition releases incompatibility between pathogen and host. All of the processes that contribute to recognition shall be referred to as "elicitation". Following elicitation, incompatibility is established, and defense reactions are expressed by the plant. The first sign of developing incompatibility in a plant cell is the oxidative burst (Apostol et al. 1989; Sutherland 1991). Protein phosphorylation (Chandra and Low 1995) and kinase-driven activation of the NADPH oxidase complex (Xing et al. 1997) mediate the formation of active oxygen compounds such as superoxide radicals O_2^-, perhydroxyl radicals HO_2, and hydrogen peroxide H_2O_2. In addition, the permeability of the plasma membrane is changed, resulting in a K^+/H^+ response. According to the ion channel defense model, signal transduction then leads to activation of the plant's defense genes. The transduction of signals depends on several enzyme activities and on the formation of second messengers (see Sect. 9.1.3.3) releasing signal cascades that also spread into cells of neighboring tissue.

A comprehensive analysis of receptor function for releasing plant defense reactions has to take account of: the nature of elicitation and receptor specificity, the steps within signal transduction leading to gene activation inside the nucleus, and the DNA nucleotide sequence of resistance genes together with their domain structure and supposed functions. The experimental systems employed for these analyses were based on the release of race-specific and active basic resistance, since both are triggered by recognition processes.

The following discussion of experiments for elucidating receptor function addresses the following issues: (1) elicitation, (2) signal transduction, (3) genetic regulation of PCD and other defense mechanisms linked to it, (4) classical genetic analysis of receptor specificity in race-specific resistance, (5) principles and techniques for identifying and isolating race-specific resistance genes, and (6) functions coded for by these genes as deduced from DNA sequencing data.

1) *Elicitation.* The first experimental system used to explore elicitor action involved non-host resistance or active basic resistance. The defense mechanisms released were from the same battery of defense responses released by race-specific resistance. However, at the time these experiments were undertaken, some of the differences between active basic and race-specific resistance were poorly understood, particularly the nature of their biochemical basis (see also Sect. 9.1.3.1), eventhough it was known that both depended on some kind of elicitation event. The elicitors employed in the first experimental system were biotic in origin, but they did not trigger hypersensitive cell death and programmed cell death, respectively. Instead they triggered synthesis of defense-related proteins in non-host plants, such as enzymes needed for the biosynthesis of phytoalexins. The employment of non-host plants for the analysis of elicitation depended on the fact that at that time only highly purified, biotic elicitors of active basic resistance were available, whereas elicitors of race-specific resistance, i.e. HR-elicitors, had not been identified. Moreover, it was realized later that an advantage of using non-host plants as model systems for investigating elicitation and release of HR (heterologous HR) is that elicitation reactions are not compromised by additional basic compatibility reactions, such as the expression of pathogenicity genes, like those observed in host plants. The experiments with non-host plants followed the expression of so-called non-hypersensitive resistance, not the release of PCD or HR. Accordingly, the elicitors employed were called non-HR elicitors (see Chap. 6). They included oligosaccharides with particular chemical structures or lipid-derived protein-containing compounds.

Special experimental systems were also set up for investigating elicitation in simplified experimental plant systems. The defense reactions of a plant are mainly limited to the infection site and its neighboring tissue. Since it is difficult to assess and manipulate biochemical reactions *in situ* in whole plants, experimental systems using plant cell suspension cultures or protoplast suspensions were developed (Mieth et al. 1986). Although the experimental conditions may differ from those in whole plants, elicitors were shown to release defense reactions under these conditions, and the biochemical reactions that were triggered were easily monitored.

An elicitor of *Phytophthora megasperma* f.sp. *glycinea* (*Pmg*, recently renamed *Phytophthora sojae*) was shown to release synthesis of furanocoumarin phytoalexins and expression of further defense reactions when added to cell cultures of the non-host plant parsley (*Petroselinum crispum*) (Rohwer et al. 1987; Parker et al. 1988, 1991). The elicitor proved to be a glucoprotein of 42 kDa which was obtained from partial acid hydrolysates of *Pmg* cell walls. This glucoprotein was also found in the culture medium of *Pmg* and was shown to be liberated from zoospores. After adding the glucoprotein to suspensions of protoplasts or microsomes of parsley, the elicitor was bound by its protein residue to specific receptors in the plasma membrane. The eliciting activity of the glucoprotein could be inactivated by treatment with pronase or trypsin, but was not affected by removal of its glucosyl residue.

By contrast, soybean (*Glycine max*), a host for *Pmg*, did not bind the purified glucoprotein elicitor to its protoplasts, as the latter apparently lack specific binding sites for the glucoprotein's protein residue. (If the glucoprotein bound to soybean, thus releasing defense reactions, so that the plant exhibited active basic resistance or basic incompatibility, then soybean would not be a host plant for *Pmg*.)

However, the partial acid hydrolysate of *Pmg* cell walls did release defense reactions in soybean, inducing in its cotyledons, or in cell cultures, synthesis of the phytoalexins pterocarpan and glyceollin (Sharp et al. 1984). This is because the acid cell wall hydrolysate of *Pmg* contained a second elicitor sensitive to deglycosylation but insensitive to treatment with pronase (Parker et al. 1988). Additional experiments showed that this elicitor was a heptasaccharide or heptaglucan with a highly specific structure. Like the elicitor for parsley, it was liberated from the β-glucans of the *Pmg* cell wall by partial acid hydrolysis. Most importantly, the heptaglucan non-HR elicitor released defense reactions in cell cultures of race-specific resistant as well as susceptible soybean cultivars. Therefore the heptaglucan was not an elicitor of race-specific resistance. Since the heptaglucan elicitor was a product of acid hydrolysis of *Pmg* cell walls, it should be considered as a biotic, but nevertheless artificial, elicitor with no significance *in vivo*. Although these experiments did not show elicitation of race-specific resistance, they proved beyond all doubt that plant cell cultures and protoplast suspensions can serve as suitable model systems for demonstrating the release of highly specific defense reactions that correspond closely to those observed in intact plants.

Further investigation of the *Pmg* glucoprotein elicitor showed that the glucoprotein can be digested by an endopeptidase to an oligopeptide of 13 amino acids, called Pep-13, which still exhibits the same eliciting activity as the undigested glucoprotein (Nürnberger et al. 1994, 1995). A minimum of 11 amino acids of specified sequence within Pep-13 are required for elicitation. ^{125}I-labeled Pep-13 binds specifically to a 91 kDa protein of a solubilized membrane preparation of parsley. Pep-13 binding seems to be species-specific, since it could not be detected in analogous membrane preparations of carrot, another umbellifer, soybean or *Arabidopsis*. The 91 kDa membrane protein has no disulfide bridges to other protein subunits and appears to be the receptor for the Pep-13 oligopeptide. Its specific binding causes elicitation of several plant defense reactions initiated by the influx of H^+ and Ca^{2+} and the efflux of K^+ and Cl^- ions (K^+/H^+ response) as well as the formation of active oxygen (oxidative burst). These early responses lead finally to reactions that include the biosynthesis of ethylene and the activation of genes coding for defense-related proteins such as those involved in the synthesis of phytoalexins.

2) *Signal transduction*: If elicitors of active basic resistance activate expression of genes from the same battery of defense genes that are involved in race-specific resistance then some type of signaling should take place between elicitation and the final expression of defense genes. Several steps within this transduction chain were mentioned in the discussion of the ion channel defense model for release of race-specific resistance (see Sect. 9.1.3.3 and *Fig.* 9). However, it is not known if the signal transduction chains that release race specific resistance and those that release active basic resistance are the same, or if they are only functionally equivalent.

Early investigations of the biochemical nature of the individual steps within the signal transduction chain leading to active basic resistance were undertaken in the non-host system employing parsley cell cultures and the glucoprotein elicitor of *Pmg* (see above). Results of these experiments corroborated the ion channel defense model (Sect. 9.1.3.3): After addition of the glucoprotein elicitor from *Pmg* an ion flux was observed through the plasma membranes of parsley cells. The ion flux began 2 – 4 min after addition of the elicitor and lasted for 3-4 h. The K^+/H^+ response was followed by the synthesis of furanocoumarin (Parker et al. 1988; Scheel et al. 1991). However, these experiments did not prove a direct connection between ion fluxes and the expression of the defense reaction, since experimental alterations of ion fluxes in the absence of elicitor did not release furanocoumarin synthesis.

Substances with other chemical structures also exhibited elicitor activity, inducing ion fluxes through the plasma membrane (Scheel et al. 1991). However, there were quantitative differences observed among the elicitors chitosan, amphotericin B and digitonin in influencing the transport of different ions. There were also differences among the several defense reactions released. Thus, various phenotypes released by elicitation may be caused by quantitative differences in ion fluxes. However, as stated in the previous paragraph, the effect of ion flux alterations on expression of defense reactions should be observed in the absence of elicitors! The phenotypic differences could have a quite different origin.

3) *Genetic regulation of programmed cell death and other defense mechanisms linked to it:* The expression of PCD can be easily monitored macroscopically. Accordingly, plant mutants were screened for modifications in the expression of PCD and subsequent activation of defense genes in neighboring cells in the absence of pathogens, with the objective of finding a system in which the reaction chains following elicitation could be broken down into discrete single steps. Such mutants were called lesion mimic mutants.

Two groups, Ausubel et al. and Dangl and Ryals, chose the crucifer *Arabidopsis thaliana* as a model system for investigating the genetic regulation of PCD. *Arabidopsis* is particularly suitable for this purpose because its life cycle is completed in 6–8 weeks, its genetics has been studied very thoroughly, and its genome, with a haploid DNA content of 130 to 140 megabase pairs, is very small compared with that of other flowering plants, facilitating the use of molecular genetic techniques. So far the majority of experiments involved *Arabidopsis* interacting with phytopathogenic bacteria. However, there is no evidence to suggest that race-specific resistance against bacteria differs from that against phytopathogenic fungi.

Ausubel and his group investigated the triggering of PCD in a homologous host-pathogen system, i.e. on the level of race-specific resistance. Seeds of *A. thaliana* were treated with the mutagen EMS (ethylmethansulfonate) and variant plants were recovered that very rapidly expressed PCD following infiltration of leaves with either virulent or avirulent *Pseudomonas syringae* pv. *maculicola* ES 4326 (Dong et al. 1991; Greenberg et al. 1993, 1994). Mutations causing this phenotype were called *acd* mutations (*accelerated cell death*). PCD was distinguished from normal cell death due to senescence by: 1. the accompanying expression of defense responses, such as the leakage of ions out of the plant cells, 2. the synthesis of phytoalexins, hydrolytic enzymes and PR-proteins, 3. the formation of callose in cell walls, and 4. the development of induced resistance (SAR, see also Chap. 11) in the plant. All *acd* mutations were recessive, and belonged to two complementation groups, *acd1* and *acd2*, that segregated independently in crosses. Their phenotypes were very similar, and reminiscent of the recessive *mlo* mutants of barley, which exhibit HR against all pathogen races of *Erysiphe graminis* f.sp *hordei* and allow, even in absence of pathogens, the expression of spontaneous HR (Jörgensen 1992; see below (4)).

Only mutants of complementation group *acd2*, for which a detailed analysis has been performed (Greenberg et al. 1994), will be dealt with here. The *acd2* locus maps on chromosome 4 of *A. thaliana*, close to *apetala2* (*ap2*), in a region harboring other genes participating in the response to pathogen attack, such as resistance to *Pseudomonas syringae* and synthesis of the phytoalexin camalexin. A characteristic of *acd2* mutants is the spontaneous occurrence of necrotic spots, or HR, on leaves in the absence of pathogens. Like HR triggered by an avirulent pathogen, the occurrence of these necrotic spots on *acd2* mutant plants is accompanied by accumulation of transcripts from genes for defense-related proteins, e.g., enzymes for the synthesis of camalexin, and for modifications of the plant cell wall. Shortly after the appearance of the necrotic spots, *acd2* plants develop induced resistance (SAR, see Chap. 11). There is a systemic accumulation of transcripts for several defense reactions, especially salicylic acid synthesis. Apparently, the HR does not solely depend on triggering by an infecting pathogen, but is also genetically determined by *acd2* and is released spontaneously without an external trigger. This leads to PCD and, via further signal transduction cascades, to the release of several different defense responses, which may be triggered coordinately by one particular signal.

Additional insight into the plant genes that regulate PCD came from the observation that the HR in *acd2-2* mutant plants infiltrated with virulent *Pseudomonas syringae* show cell wall modifications, accumulation of defense related transcripts, salicylic acid, and antimicrobial compounds. Bacterial growth is also reduced like that resulting from avirulent pathogens in wild type plants. HR may develop on the cotyledons of *acd2-2* plants spontaneously and, as mentioned above, can be triggered on adult leaves in the absence of any bacteria by wounding or mechanical stress. Furthermore, HR induced by virulent or avirulent bacteria, spreads in *acd2-2* mutant plants within a few days over the whole leaf, even in the absence of any spreading of the bacteria from the site of infection. Such a phenotype is never observed on wild type plants infected with avirulent bacteria. The spreading of PCD also occurs after spontaneous HR induction or by wounding or mechanical stress. This suggested that in *acd2* mutant plants PCD resulted from a loss of regulation which depends on an active contribution by the plant cell, as the mutation allows the spreading of necrosis, possibly by spreading the signals which trigger it. Thus, the regulatory mechanism active in wild type plants seems to be a negative one. Correspondingly, *acd2* mutations are recessive and are not expressed in heterozygous *acd2/Acd2* plants, the wild type *Acd2* product acting as a negative regulatory product.

Direct evidence for a metabolically dependent, plant-determined, cell-killing program may be seen in the action of the harpin$_{Pss}$ protein. This protein is produced by the phytopathogenic bacterium *Pseudomonas syringae* pv. *syringae* and is an elicitor of HR (He et al. 1993). The protein, which has been purified, releases HR in leaves after infiltration. However, the release of HR fails if the harpin protein is infiltrated together with metabolic inhibitors, such as amanitin (blocks eukaryotic polymerase II), cycloheximide (blocks the function of 80S ribosomes), vanadate (blocks ATPases, and phosphatases), or lanthanum (blocks Ca^{2+} ion channels). PCD also occurs in animal cells (Horvitz et al. 1982; Oppenheim et al. 1990; Hockenbery et al. 1993; Raff 1991, 1993; Veis et al. 1993), where it is known as apoptosis. In the nematode *Caenorhabditis elegans* and in mice, apoptosis, as in plants, is negatively regulated. PCD occurs in vertebrates during development of the nervous system.

Very similar results and conclusions were reached by the Dangl and Ryals group (Dietrich et al. 1994) who analyzed PCD expressed in active basic resistance. In their system, *Arabidopsis* variants, isolated after insertional mutagenesis, that spontaneously formed necrotic spots, were called *lesion simulating disease mutants, lsd.* Mutant phenotypes were expressed either always or only conditionally, under short- or long-day conditions. Short-day conditional *lsd* mutant plants could be grown in long days, the permissive conditions without expressing the mutant phenotype. Plants transferred from permissive long-days to short-days expressed spontaneous HR. All of the *lsd* mutants were monogenically determined and could be classified into five complementation groups, each exhibiting under non-permissive or short-day conditions the same release of defense gene expression as observed in the *acd* mutants (see above). Some mutants were recessive (*lsd1, lsd3* and *lsd5*), and others dominant (*lsd2* and *lsd4*). Mutants *lsd2, lsd3, lsd4* and *lsd5* were called initiation mutants and were resistant to virulent pathogen isolates, and the recessive *lsd1* mutant was called a feedback mutant (see below). Since the expression of PCD in *lsd* mutants is independent of any genes for race-specific resistance it has to be classified as active basic resistance.

The spontaneously induced but locally restricted necrotic HR spots of the initiation mutants cannot be enlarged by pathogen infection or by applying substances, such as salicylic acid, which trigger the onset of induced resistance (SAR, see Chap. 11). (In the recessive *lsd3* and *lsd5* mutants, formation of HR spots may be due to loss of a controlling function, and in the dominant *lsd2* and *lsd4* mutants it may be the result of constitutive gene expression.)

The phenotype of the dominant initiation mutants *lsd2* and *lsd4* is lost in *NahG* transgenic plants. These plants have low levels of intracellular salicylic acid, since the enzyme salicyl hydroxylase, encoded by the *NahG* transgene, instantly breaks down salicylic acid to catechol after it has been synthesized via phenyl propanoid metabolism. Apparently, an elevated level of salicylic acid is required for expression of plant defense genes (Hunt et al. 1996). This should also hold true for the expression of induced resistance (SAR). Thus, salicylic acid represents an important component of the signal transduction chains bound to PCD and SAR (Delaney et al. 1994, also see Chap. 11; *Fig. 15*). For example, the phytopathogenic bacterium, *Pseudomonas syringae* pv. *tabaci* carrying avirulence gene *AvrRpt2* is avirulent on *A. thaliana* ecotype Columbia (Col-0) *Rpt2*, but is virulent on the same *Rpt2* resistant ecotype carrying the *NahG* transgene. This indicates that a low level of intracellular salicylic acid prevents expression of gene-for-gene resistance expression. Similarly, the Wela and Emwa races of *Peronospora parasitica* are avirulent on the same *A. thaliana* ecotype Columbia (Col-0), but highly virulent on the *NahG* transgeneic *A. thaliana* Col-0 (Delaney et al. 1994).

The recessive or feedback mutant, also referred to as a propagation mutant, behaves differently from the initiation mutants. It has lost control over the spread of PCD, as observed with the *acd2* mutants (see above): After infection with *Pseudomonas syringae* or *Peronospora parasitica,* or treatment with chemical inducers of SAR, spreading of PCD is triggered over the whole leaf. This so-called "runaway cell death" belongs by definition to active basic resistance since no race-specific resistance genes are involved. This seems to be caused by cascades of propagating signals, among which the superoxide anion O_2^- plays a central role. Accordingly, pressure injection of an O_2^- generating system (xanthine – xanthine oxidase) alone into the intercellular space of leaves of *lsd1 - Arabidopsis* induces runaway cell death (Jabs et al. 1996). However, mechanical injury of *lsd1* mutant plants does not trigger spreading of hypersensitive cell death, although in the recessive *acd2* mutants both infection and mechanical injury trigger the spreading of necrosis. PCD mutations have also been identified in other plants besides *Arabidopsis*, for example, in leaves of maize and rice plants and in roots of soybean.

To summarize: the release and extent of PCD may be modified by several different single step mutations. These mutants have phenotypes that are identical, or very similar to the expression of hypersensitivity at the levels of active basic resistance, and race-specific resistance and may be expressed in the absence of pathogen attack. This, and the fact that they were derived from quite different selection systems, strongly suggests that specific recognition of the elicitor and regulation of HR are closely associated processes. However, neither the *acd* and *lsd* nor the *mlo* mutants have been mapped to genes conferring race-specific resistance which suggests that not all lesion mimic mutants have a direct evolutionary link to resistance genes (Richter and Ronald 2000).

As discussed in Chaps. 5 and 9, Sec. 9.1.3.1, hypersensitive cell death (HR, PCD) appears to be linked with the induction of defense gene expression in neighboring cells not afflicted by HR. In rare cases however, after gene-for-gene interaction, either expression of HR or of defense genes was observed, but not both reactions together. This suggested that HR may not be essential for gene-for-gene triggered expression of defense genes. If so, one might expect to find, for example, plant mutants in which only defense genes are expressed, following gene-for-gene interaction, but no hypersensitve cell death. Indeed, such HR mutant phenotypes have been identified. After screening about 11,800 plants from EMS-mutagenized M_2 seeds of carrying the race-specific resistance gene *RPS2*, 13 plants were obtained exhibiting an HR phenotype when infected with *Pseudomonas syringae* pv. *glycinea* (*Psg*) carrying the avirulence gene on a plasmid. In these interactions the plant mutants, after vacuum infiltration of *Psg* into the leaves, developed three different HR⁻ tissue collapse phenotypes after 24 and 40 hours, namely: intermediate or weak, delayed, or no HR cell collapse at all (Yu et al. 2000, 1998). The mutants shared the following properties: After crossing with unmutated plants the determinant for each of the different HR phenotypes segregated as a single, recessive gene. The F_1 progeny developed wild type HR. However, the defective HR expression of the homozygous recessives did not compromise the expression of disease defense genes triggered by the *RPS2 - AvrRpt2* gene pair thereby preserving the incompatible interaction phenotype. Furthermore, none of the mutants exhibited the lesion mimic phenotype like the *acd* and *lsd* mutations (see above in this chapter). One of the *RPS2* resistant HR⁻ mutants, carrying the *dnd1* (defense *no* death) allele, was characterized in more detail (Yu et al. 1998). The wild type gene *Dnd1* mapped on an arm of chromosome 5 where no resistance genes have been found so far. Growth of avirulent *Psg* was severely inhibited in *RPS2 dnd1* mutants. Interestingly, even virulent *Psg*, i.e. in the absence of a gene-for-gene interaction, showed a 10 to 100-fold growth inhibition in *RPS2 dnd1* plants, as typically observed in plants exhibiting induced resistance (SAR; see Chap. 11). Accordingly, *dnd1* plants inoculated with different pathogens like tobacco ringspot virus, *Xanthomonas campestris* pv. *campestris*, or *X. campestris* pv. *raphani*, developed weaker disease symptoms and *Peronospora parasitica* produced three fold less spores on *dnd1* mutant plants. In the absence of infecting pathogens the *dnd1* mutants showed constitutive expression of PR genes (PR-1 mRNA, β-glucanase) and an increased level of salicylic acid and glucoside-conjugated salicylates suggesting that gene *Dnd1* is involved in PR gene regulation. If *Rps2 dnd1* plants are infected by *Psg* carrying the corresponding avirulence gene *AvrRpt2* the resulting gene-for-gene interaction boosted the growth inhibition by a factor of 10^3 to 10^4. These results indicate a gene-for-gene triggered defense gene expression despite the virtual elimination of HR cell death demonstrating the independence of defense gene expression on signal substances produced by cells sustaining HR. Rather, the signal transduction following gene-for-gene recognition has two branches, one leading to HR, the other to the expression of defense genes (see also Bowling et al. 1997). Thus HR only strengthens the effect of defense gene expression.

4) *Classical genetic analysis of receptor specificity of race-specific resistance genes:* Classical genetic analysis of race-specific resistance relies on the specificity of elicitation, and signal transduction to produce recognizable phenotypes.

According to the gene-for-gene hypothesis, the membrane-bound race-specific resistance factor and its receptor are determined in the plant by a single gene. The same holds true for the avirulence factor of the pathogen. Whereas most other genes exhibit mutation rates of about 10^{-6}, avirulence genes of phytopathogens may have mutation rates that are much higher, by a factor of $10^2 - 10^3$, and in some cases mutation rates of 1% have been observed for defective avirulence, i.e., specific virulence. Even if the mutation rates of most avirulence genes do not reach this magnitude, the high reproduction rates among pathogens result in substantial absolute numbers of specific virulent mutants in pathogen populations. Thus, race-specific resistance is a rather unstable genetic trait, because any mutational event in a pathogen's avirulence gene, or its loss by deletion, leads to defective avirulence, and escape of the pathogen from recognition by the plant's resistance gene product thus effecting expression of specific virulence.

During evolution, both susceptible and race-specific resistant plants should have been prone to extinction by their homologous pathogens, avirulent wild type pathogens infecting susceptible host plants, and specific virulent mutants infecting race-specific resistant ones. While this may have occurred for species that are now extinct this has obviously not occurred for modern higher plants. Evidently they were able to escape eradication by continually developing new resistance specificities for recognizing the products of homologous pathogens as avirulence factors. Of course, this defense strategy requires that mutations to race-specific resistance occur in plants at about the same frequency as mutations to specific virulence occur in pathogens. However, experiments undertaken to verify this supposition demonstrated that no novel resistance specificities, either spontaneous or induced, could be obtained in susceptible plants except those already known. In barley, the only novel resistance type isolated in such experiments was a mutation in the *Mlo* locus to resistance against powdery mildew, *Erysiphe graminis* f.sp. *hordei*, causing HR (Jörgensen 1992).

Surprisingly, this hypersensitive reaction was expressed against all available races of *E. graminis* f.sp. *hordei*. Thus, the isolated mutant did not carry a novel race specificity, rather it lost any race specificity. Furthermore, *mlo* mutant barley plants exhibited, even in the absence of pathogens, a more or less pronounced propensity to express necrotic HR lesions on their leaves, and they formed callose appositions at the cell walls of the layer below the necrotic spot. In the absence of pathogens, the formation of necrotic spots could be triggered with high efficiency by rubbing the leaves with carborundum. Since the *mlo* mutation does not exhibit race-specificity, the gene is distinct from those associated with the gene-for-gene type of race-specific resistance and was classified as a lesion mimic mutant (see above). All *mlo* mutants so far isolated belong to a family of recessive and non-complementing alleles mapping within the *Mlo* gene to a chromosomal region harboring none of the dominant race-specific mildew resistance genes of barley. Thus, it is unlikely that the wild type gene *Mlo* encodes a receptor protein recognizing avirulence product(s). Rather, the locus seems to determine a regulatory function. Significantly, the phenotypes of *mlo* mutants of barley are very similar to those of the signal transduction-related *acd* mutants of *Arabidopsis* discussed above.

Classical genetic analysis of race-specific resistance also provided some clues as to the nature of signal transduction following elicitation. The failure to obtain mutants of plants with novel resistance specificities was startling because the appearance of novel mutant types at frequencies as high as those observed in the pathogens was expected. Thus, it appeared that novel resistance specificities could not be produced by normal spontaneous or induced mutations. The outcome of experiments to induce mutations in race-specific resistant plants was similar. Race-specific resistant barley cultivars carrying an allele for resistance at the *Mla* locus were mutagenized and screened for new types of resistance mutants. The only mutants found in high frequencies had the same race specificity but reduced resistance, i.e., increased susceptibility to attack by *Erysiphe graminis* f.sp. *hordei*, combined with somewhat altered infection phenotypes. Thus novel, or qualitatively altered, resistance specificities were not obtained either in a race-specific resistant or a susceptible cultivar. Only susceptible revertants or mutants with partly reduced resistance but unchanged specificity were found. It was suggested that these quantitative effects were due to mutation in genes required for the function of the resistance genes. For example, they may be involved in signal transduction reaction(s) following recognition of the avirulence product and the expression of defense reactions.

The first evidence for the existence of genes required for functions following recognition in race-specific resistance came from these same experiments on barley resistant against powdery mildew, *E. graminis* f.sp. *hordei*. After mutagen treatment of seeds of *Ml-a12* resistant plants, 25 defective mutants were isolated with various degrees of reduced resistance but the same race specificity (Torp and Jörgensen 1986; Jörgensen 1988). The mutagen treatments were X rays, EMS, or NaN$_3$. Treated seeds were sown to raise M$_1$ plants which were selfed and sown to raise M$_2$ plants which were inoculated with *E. graminis* f.sp. *hordei* carrying the corresponding avirulence gene and screened for defective resistance. Twenty five defective mutants were isolated and backcrossed with *ml-a12*, the susceptible wild type plant. Only susceptible plants were found among the progenies in the majority of back crosses, indicating that a mutation to susceptibility had occurred in the *Ml-a12* gene. However, in about 10 % of the crosses, resistant segregants were also recovered. In these cases, the mutation for more or less defective resistance could not have occurred in the *Ml-a12* locus, but rather in another gene required for *Ml-a12* function. In the back crosses the *Ml-a12* resistance gene segregated from the other gene which prevented full expression of resistance. This observation was the first unequivocal evidence that supplementary genes are required for releasing race specific resistance. It seemed reasonable that these genes might be involved in signal transduction. In the same system two such supplementary genes were later identified, *Rar1* and *Rar2* (originally called by Freialdenhoven et al. 1994 *Nar-1* and *Nar-2*).

Following mutagen treatment of tomato carrying the race-specific resistance gene *Cf9*, eight new recessive mutations were isolated: four with complete and four with extensive loss of resistance against *Cladosporium fulvum* carrying the wild type gene *Avr9* (Hammond-Kosack et al. 1994). Two of the four mutants with reduced resistance, *rcr-1* and *rcr-2* which are inherited independently of each other, map at loci unlinked to *Cf9*. The infection phenotype of the four fully susceptible mutants was different from that of a *Cf0* plant, a cultivar with no known race-specific resistance to *C. fulvum*. The effects of the wild type *Rcr* genes (required for *Cladosporium resistance*) on *Cf9* expression remain unknown, but they are believed to involve other *R* gene functions than those that encode specificity (see (6).

Since it was not possible to generate variants of plants with altered race-specific resistance by mutagen treatment one may ask how such resistance mutations could arise at sufficient frequencies even in small populations to guarantee survival in the presence of homologous pathogens. The observation that genetic determinants for race-specific resistance map as closely linked clusters or as multiple alleles of a single gene suggested that novel resistance specificities might arise by recombination (Richter and Ronald 2000). The alleles of closely linked genes may recombine by crossing over, whereas multiple alleles in general show no recombination. However, closely linked alles may recombine at frequencies too low to be detected. For this reason recombination is not a fool proof way of discriminating between alleles of different genes or of the same gene. However, such discrimination can be accomplished among F_2 progenies in the so-called modified allelic test (Shephered and Mayo 1972).

Clustered resistance genes have been found in the M and L loci of flax against flax rust, *Melampsora lini*, in the *Rp1* locus of maize against maize rust, *Puccinia sorghi*, and in the *Mla* and *Mlk* loci of barley against powdery mildew, *Erysiphe graminis*. Within the M locus of flax four resistance specificities map as four separate, closely linked genes, whereas in the L locus eleven different specificities were shown to be multiple alleles of a single gene (Pryor and Ellis 1993; Lawrence et al. 1994).

A comparable clustering has also been found for genes determining self-incompatibility in flowering plants (de Nettancourt 1977; Nasrallah et al. 1994). After spontaneous or induced mutagenesis, only self-compatible variants with no specificity were found. There were no mutants showing an altered specificity. Fisher (1961) suggested that in a small chromosomal region with closely linked genes or multiple alleles, new incompatibility types could arise in high frequencies by unequal crossing over, or illegitimate recombination, without requiring an increase in mutation frequency.

Some evidence that recombination events may participate in generating novel resistance specificities comes from crosses between maize cultivars resistant and susceptible to maize rust, *Puccinia sorghi*. Resistant parental cultivars – each carrying an allele of one of the closely linked genes in the *Rp* locus – were crossed with a susceptible pollen donor plant. In the F_1 generation, pathogen-susceptible progeny appeared at frequencies of $10^{-2} - 10^{-4}$. Since race-specific resistance is dominant susceptible segregants should not appear. The unexpected susceptible segregants had exchanged flanking markers, suggesting that they originated from a recombination event caused by unequal crossing over within the *Rp* locus carrying the resistance alleles (Bennetzen and Hulbert 1992; Lawrence et al. 1994; Richter and Ronald, 2000). In similar crosses involving flax plants carrying alleles for rust resistance at the L locus no crossing over was detected. However, in crosses of flax plants heterozygous for two L alleles, among 71,266 progeny plants tested, 56 had resistance phenotypes that differed from the parents (Pryor and Ellis 1993). Among the 56 segregants two classes could be distinguished: 43 were susceptible, but 13 exhibited modified and spreading resistance phenotypes which had not been observed previously in flax, and some had an altered spectrum of resistance to different rust races. This observation documents the first case of the recovery of new resistance specificities from previously existing genes. However, the possibility that the 56 segregants arose from a recombination event remains speculative since no flanking markers were available.

The nature of the 43 susceptible segregants, whether they also originated from recombination, and whether they are all of the same type is still largely unknown. Some of the segregants were tested for reversion to resistance, and in rare cases the original resistance specificity, of one of the parents reappeared. Evidently the genetic information determining the parental resistance specificity was still present in the susceptible segregants, but not in a form that allowed its expression. We may also suppose that, among the susceptible segregants, other recognition specificities were present which escaped identification, either because pathogens with the corresponding avirulence products were not present among the tester races used, or because corresponding avirulence factors did not exist. If true this would mean that it would be unlikely that all recombination events within a plant's resistance determinant would give rise to recognition specificities and receptor structures for which a corresponding avirulence product is available.– There is no evidence for generation of new specificities at resistance gene loci as a result of insertion or excision of a transposable element (Richter and Ronald 2000).

To summarize: The results discussed above suggest that recombination events within a chromosomal locus are the source of new race-specific resistance specificities. These events probably also contribute to the host plant genetic polymorphism (see Sect. 9.1.1) and may happen very frequently. However, the majority of new recognition specificities will escape "identification" by an infecting pathogen, if the latter is just not present in the tester races employed or if such avirulence products do not exist, either in this host-pathogen system or in nature.

5) *Molecular genetic methods for the identification and cloning of race-specific resistance genes.* According to the elicitor-receptor model of the gene-for-gene hypothesis a resistance gene should only encode the receptor protein that recognizes the pathogen's avirulence product. However, in the ion-channel-defense model (see Chap. 9; Sect. 9.1.3.3) the *R* gene is assumed, to encode several additional functions following recognition, such as signal transduction. This raises the question of which molecular genetic methods are most suitable for cloning genes determining race-specific resistance? (Ellis et al. 1988). Generally speaking, cloning a gene means the identification and selection of a DNA fragment containing the gene which can be multiplied in a suitable vector at will. In this form its nucleotide sequence can be determined and also the structure and function of the protein it forms and how its expression is regulated. Identification and selection may be carried out by two different approaches: the gene is

identified either by (1) its product, or phenotypic expression, or (2) on the basis of a base sequence which acts as a marker in the DNA fragment carrying the gene. The ion channel defense model suggests that the products of R genes are proteins which serve several different functions. However, at the present time so little is known about the different functions of R genes or their products that method (1) cannot be applied. However, recent studies employing "map-based cloning" and "transposon tagging" corresponding to method (2) were very successful (de Wit 1995).

In map-based cloning, the gene is localized within the DNA between two molecular markers, such as RFLPs (restriction fragment length polymorphisms) that are easily mapped. These should flank the gene as closely as possible (van den Beek et al. 1992; Jones et al. 1993; Bent et al. 1994). The region between the flanking markers can then be cloned step-by-step using "chromosome walking" and the DNA stretches they embody can be sequenced. The resistance gene, with its reading frame, should be found within the sequenced DNA stretch. Comparison of the nucleotide sequence of the gene with the sequences of other genes deposited in gene bank databases allows tentative conclusions to be made about the functions encoded by the different functional domains of the gene.

In transposon tagging, the R gene is functionally and molecularly marked by the insertion of a transposable element. A transposon is a small, autonomous DNA molecule able to insert, randomly and at low frequency, into various sites in chromosomal DNA, where it prevents the expression of the gene located at the site of insertion. Thus transposon inactivation of R gene function, in combination with molecular mapping of the transposon's insertion site, localizes the site of the R gene unambiguously (Rommens et al. 1992). This technique was used to identify and clone the Cf9 gene of tomato (Jones et al. 1994). Plants carrying the transposon-marked Cf9 resistance gene were isolated by using an ingenious selection technique. The population of transposon-mutagenized plants was crossed with susceptible Cf0 (i.e. Cf9 free) plants containing a transgene for the avirulence gene Avr9 from the tomato pathogen Cladosporium fulvum. Among the F_1 progeny, all plants which carry the transgene Avr9 together with an active Cf9 gene will die, since when both genes are expressed seedlings will undergo HR in all cells from early development on and will be killed by PCD (see also Sect. 9.1.3.1). Only hybrid seedlings carrying the Avr9 transgene together with a Cf9 gene *inactivated* by transposon insertion will survive, since no PCD is released. From such F_1 hybrid plants the Cf9 gene was isolated and cloned (Jones et al. 1994). From about 160,000 F_1 hybrid seeds 118 viable seedlings were obtained of which 37 on further testing were shown to carry the Cf9 transposon tagged gene.

6) *Functional domains in genes determining race-specific resistance.* Several R genes from five plant species have been identified and cloned so far by employing map-based cloning and transposon tagging. All of these genes are relatively large

and encode about 1000 amino acids. Although no receptor proteins have been isolated so far, the amino acid sequences of the proteins and their secondary and tertiary structures could be deduced from their nucleotide sequences. Sequence comparisons with those of known proteins have indicated different functional domains within the R genes.

The race-specific resistance gene *Cf9* of tomato against the pathogen *Cladosporium fulvum*, maps within a complex locus that also includes the R genes *Cf4* and *Cf1* (Jones et al. 1993, 1994). Gene *Cf9* encodes 863 amino acids (AA) together with a signal peptide of 23 AA which is cleaved off after its secretion. The product of *Cf9* is a glycoprotein, with a hydrophobic carboxy terminus of 76 AA which functions as a transmembrane domain anchored in the plant cell's plasma membrane. The much larger glycosylated region consists of the remaining 764 AA and is localized outside the cell membrane. This part of the molecule includes 27 imperfect leucine-rich repeats (LRR), each containing, on average, 24 AA which, in contrast to globular polypeptides, provides a regularly organized, linear protein structure which can take part in protein-protein interactions. LRRs are widespread and appear to be a motif that during evolution has been adapted to many different functions (Kobe and Deisenhofer 1994). In R gene products, LRRs are believed to recognize the avirulence factor and also to play a role in signal transduction cascades via phosphorylation that are dependent on protein-protein interaction. They are also likely to undergo recombination to generate new recognition specificities (Richter and Ronald 2000).

The *Cf2* locus has two equally large and fully functioning *Cf2* gene copies which differ by only three amino acids. Each *Cf2* gene encodes a protein of 1112 AA, including a 26 AA signal peptide active in excretion of the protein through the cell membrane (Dickinson et al. 1993; Dixon et al. 1996). Several regions within the *Cf2* DNA exhibit high homology with the sequence of the unlinked *Cf9* gene. The differences in size between the processed Cf2 and Cf9 proteins is due to a larger LRR in *Cf2* (Hammond-Kosack and Jones 1997).

By 1966 DNA sequences of ten different R genes had been published (Hammond-Kosack and Jones 1997). Five encode race-specific resistance against phytopathogenic fungi (*Cf9*, *Cf2*, *and Cf4* in the tomato/*C. fulvum* system, *L6* in flax/ *M. lini*, and *RPP5* in *Arabidopsis*/*Peronospora parasitica*), four encode resistance against phytopathogenic bacteria (*RPS2* and *RPM1* in *Arabidopsis*/*Pseudomonas syringae* pv. *tomato* and *Pseudomonas syringae* pv. *maculicola*, *Xa-21* in rice/ *Xanthomonas oryzae* pv. *oryzae*, and *Pto* in tomato/*Pseudomonas syringae* pv. *tomato*), and *N* in tobacco for resistance against the mosaic virus TMV. DNA sequencing of these race-specific resistance genes and the amino acid sequences derived from them revealed a rather surprising result: The resistance gene products identified so far belong to two different classes of receptors, those located extracellularly, harboring one "transmembrane domain", consisting of hydrophobic amino acids, and those located in the cytoplasm lacking such a domain. The differences in the locations of resistance proteins most probably mirror the site of recognition between avirulence factor and resistance protein. However, their location appears to be independent of whether the pathogen is fungal, bacterial or viral. Finding intracytoplasmic receptors was unexpected since recognition was thought to occur only at the outer cell surface (see *Fig. 10*).

Resistance proteins Cf9, Cf2, and Cf4 in tomato (conferring resistance to *Cladosporium fulvum*) have their transmembrane domains at the carboxy terminus of the molecule, so that their LRRs for elicitor recognition and their amino termini are located extracellularly.– The receptor protein Xa21 in rice (conferring resistance against the pathogenic bacterium *Xanthomonas oryzae*) is somewhat similar. However, the transmembrane domain of the protein is in the middle of the molecule where it is anchored within the plasma membrane. Thus the molecule's carboxy terminus is located within the cytoplasm and the amino terminus is extracellular (see *Fig. 10*). This class of receptor proteins fits the classic model for recognition between receptor and elicitor.

The three types of intracytoplasmic resistance gene products shown in *Figure 10* (Pto, Prf, N, L6, RPP5, RPS2, and RPM2) are the products of their respective genes (see next to the last paragraph). Gene *Prf* is closely linked to *Pto* of tomato and also involved in expression of resistance (Martin et al. 1996, Salmeron et al. 1994). Except for Pto, the resistance gene products Prf, N, L6, RPP5, RPS2, and RPM2, shown in *Fig. 10*, also contain LRRs but they lack a transmembrane domain and are therefore located within the cell's cytoplasm. However, like the membrane-bound proteins, they contain LRRs and thus have the potential for recognition and signal transduction. Therefore, the highly specific elicitor receptor recognition most likely takes place within the cytoplasm. How the eli-

---

Figure 10. Different receptor proteins of race-specific resistance genes and their functional domains, located either extracellular or intracellular (*columns*) as derived from nucleotide sequences of the cloned *R* genes. *Columns* correspond to assumed sizes of the respective native proteins, the *numbers* next to the names of resistance genes apply to the numbers of amino acids. From *left to right*: Race-specific resistance gene *Xa-21* in the system rice/*Xanthomonas oryzae* p.v. *oryzae*, genes *Cf9*, *Cf2* and *Cf4* in the system tomato/*Cladosporium fulvum*, genes *Pto* and *Prf* in the system tomato/*Pseudomonas syringae* p.v. *tomato*, gene *L6* in the system flax/*Melampsora lini*, gene *N* in the system tobacco/TMV, gene *RPS2* in the system *Arabidopsis*/*Pseudomonas syringae* p.v. *tomato* and gene *RPM1* in the system *Arabidopsis*/*Ps. syringae* p.v. *maculicola*. *Double arrows* with question-marks, speculative protein-protein interactions (Hammond-Kosak and Jones 1997), not discussed in the text; *N*, the amino terminus; *C*, carboxyl terminus. Different domains within a resistance protein: *K*, kinase domain; the serine-threonine kinases serve in signal transduction. *LRR*, leucine-rich repeat domain binds the avirulence factor, responsible for recognition between elicitor and race-specific receptor, and is involved in signal transduction. *LZ*, leucine zipper of homologous di- or oligomerization as well as heterologous dimerization. *NBS*, domain participating in activation of G-proteins and phosphate kinases. *TIR*, domain with homlogy to the cytoplasmic domain of *Drosophila* Toll and to the human interleukin-1 receptor. *TMD*, transmembrane domain used in anchorage of the resistance protein in the plasma membrane.

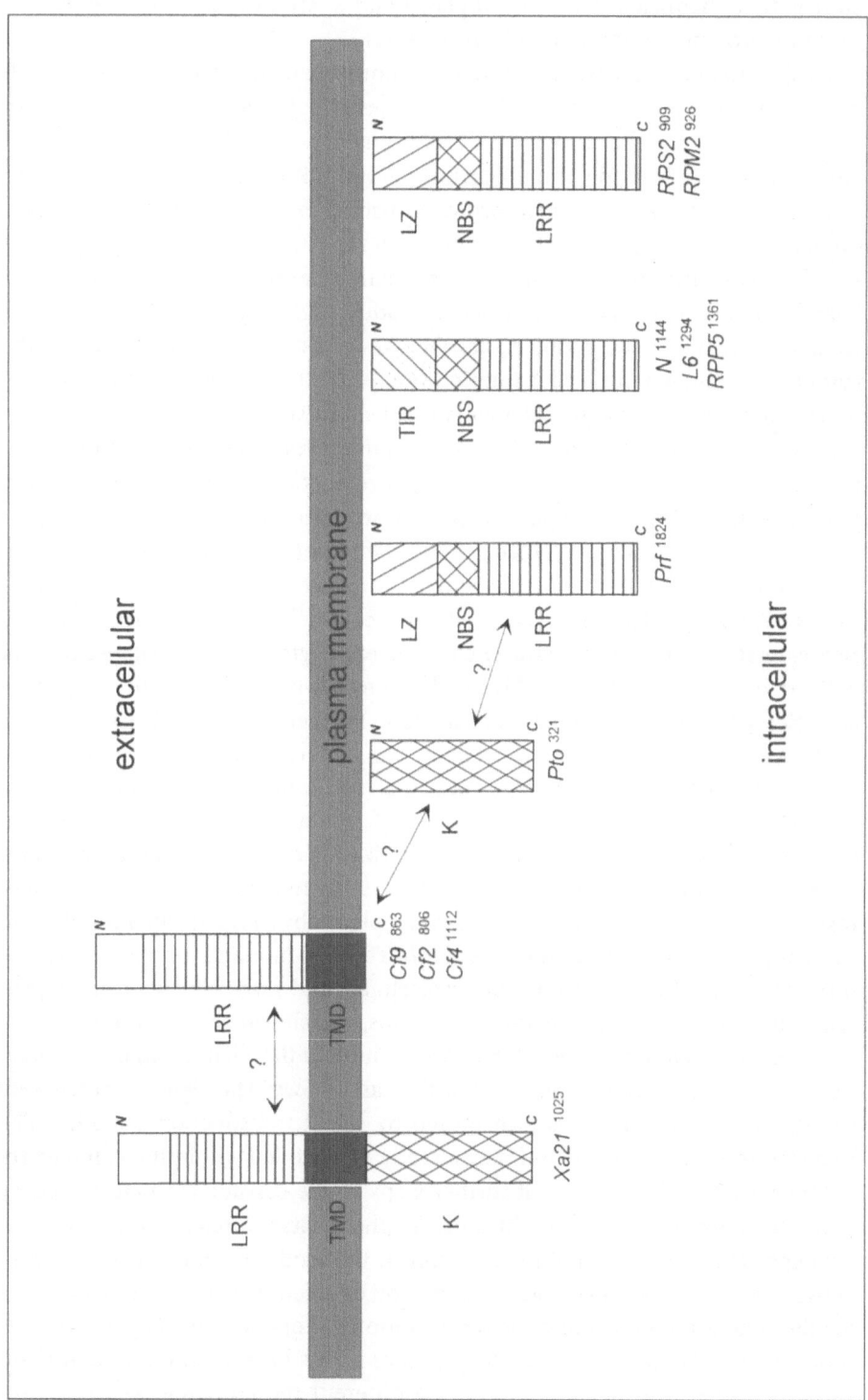

citor is first channeled into the cytoplasm and is afterwards recognized by the receptor protein is a topic of current research.

DNA domains with particular sequence motifs were detected in most if not all of the ten genes so far sequenced. This suggests that at least some of the functional processes for elicitation and signal transduction are common to race-specific resistance genes (de Wit 1995; Boyes et al. 1996; Hammond-Kosack and Jones 1997). In all resistance gene proteins, distinct extracellular and intracellular domains containing glycosylated leucine-rich repeats (LRR) are present (Bent et al. 1994). The LRRs are believed to interact with other proteins, carbohydrates or ligands, such as elicitors and avirulence products. LRRs are also engaged in the transduction of the recognition signal. Only the resistance proteins located in the cytoplasm contain a nucleotide binding site (NBS) that possibly activates G proteins and phosphate kinases, elements involved in signal transduction. By contrast, the majority of extracytoplasmic resistance proteins within the *Cf* gene family of tomato, such as Cf9, Cf4, and Cf2, have no kinase domains but probably interact with a separate cytoplasmic protein with kinase activity (K). A comparable situation exists for resistance gene *Pto* of tomato, coding exclusively for an intracytoplasmic kinase interacting with a protein coded for by gene *Prf* containing a LRR domain. The partly extracellular race-specific resistance gene *Xa-21* of rice against *Xanthomonas oryzae* encodes an extracytoplasmic resistance protein with its own intracytoplasmic kinase domain. All kinase and phosphorylating activities participate in signal transduction. Furthermore, avirulence products Avr9, Avr4, and Avr2 seem to interact with the resistance proteins of genes *Cf9*, *Cf4* and *Cf2*, either directly or in association with a binding protein. Also, resistance genes *RPS2*, *RPM2* and gene *Prf* contain a leucine zipper domain (LZ) facilitating homologous di- and oligomerization or heterologous dimerization of proteins (Bent et al. 1994; Mindrinos et al. 1994). In the intracellular protein of resistance gene *L6* of flax the LZ domain is replaced by a region with homology to the cytoplasmic domains of the *Drosophila* Toll and human interleukin-1 receptors (TRS). In *Fig. 10* examples of extracellular and intracellular receptor proteins, and their different functional domains, are shown schematically.

The events that take place between recognition of the elicitor and final expression of defense genes may be summarized as follows: The signal generated by recognition of the elicitor is transmitted by signal transduction cascades. The first step in signal transduction may already be encoded within the complex resistance gene itself, whereas all further steps in the cascade are determined by separate genes encoding protein kinases, phosphatases, phospholipases or G-proteins. This results in both expression of the oxidative burst (generation of active oxygen compounds) and the activation of defense genes in the cell nucleus via the generation of cytoplasmic transcription factors (see also *Fig. 9*). Furthermore, all results obtained so far suggest that plants have several different pathways for signal transduction, employing different receptors and assisting proteins, rather than a single universal pathway. Plant defense strategies also seem to be directed against attack by any kind of pathogen, and no special defense mechanisms are employed for defense against specific pathogens. Apparently, a single pathogen sets in motion several signal transduction pathways, which release different defense reactions, such as deposition of callose or lignin, synthesis of phytoalexins and salicylic acid, activation of pathogenesis related proteins (PR-proteins), or hypersensitive cell death (HR or PCD). This strategy guarantees the simultaneous expression of a cocktail of several different defense reactions against one attacking pathogen.

The protein structures and functions derived from sequencing R genes and other genes engaged in signal transduction are consistent with the ion channel defense model (see Sect. 9.1.3.3). However, we may ask if the results discussed above can be reconciled with the gene-for-gene hypothesis originally proposed by Flor? The answer is yes, if one considers the same facts that Flor did; that is, the variability of different cultivars with respect to their interaction with one particular pathogen avirulence factor. In other words, the specificity required for recognition between an avirulence product and a receptor protein depends on one avirulence gene in the pathogen and one resistance gene in the plant. The signal transduction cascades following in the plant are not involved in the recognition reaction and are, at least in principle, common to any race-specific recognition reaction.

Reviews

Atkinson M.M. (1993): Molecular mechanisms of pathogen recognition by plants. Advances in Plant Pathology 10: 35–64

Bennetzen J.L., Hulbert S.H. (1992): Extramarital sex amongst the beets – Organization, instability and evolution of plant disease resistance genes. Plant Mol.Biol. 20: 575–580

Bent A.F. (1996): Plant disease resistance genes: Function meets structure. Plant Cell 8: 1757–1771

Beynon J.L. (1997): Molecular genetics of disease resistance: an end to the "gene-for-gene" concept? In: Crute I.R., Holub E.B., Burdon J.J. (eds.): The Gene-for-Gene Relationship in Plant-Parasitic Interactions. CAB International, Oxon UK, New York NY. 359–377

Boller T (1995): Chemoreception of microbial signals in plant cells. Annu.Rev.Plant Physiol.Plant Mol.Biol. 46: 189–214

Boyes D.C., McDowell J.M., Dangl J.L. (1996): Many roads lead to resistance. Curr.Biol. 6: 634–637

Brettell R.I.S., Pryor A.J. (1986): Molecular approaches to plant and pathogen genes. In: Blonstein A.D., King P.J. (eds.): Plant Gene Research: A Genetic Approach to Plant Biochemistry. Springer-Verlag, Vienna, New York. 233–246

Dangl J.L. (1992): The major histocompatibility complex à la carte: are there analogies to plant disease resistance genes on the menu? Plant J. 2: 3–11

Dangl J.L. (1993): The emergence of *Arabidopsis thaliana* as a model for plant-pathogen interactions. Advances in Plant Pathology 10: 127–156

Dangl J.L. (1995): Pièce de Résistance: Novel class of plant disease resistance genes. Cell 80: 363–366

Dangl J.L., Dietrich R.A., Richberg M.H. (1996): Death Don't Have No Mercy: Cell Death Programs in Plant-Microbe Interactions. Plant Cell 8: 1793–1804

De Wit P.J.G.M. (1995): Fungal avirulence genes and plant resistance genes: Unravelling the molecular basis of gene-for-gene interaction. Adv.Bot.Res. 21: 147–185

Ebel J., Grisebach H. (1988): Defense strategies of soybean against the fungus *Phytophthora megasperma* f.sp. *glycinea*: a molecular analysis. Trends Biochem.Sci. **13**: 23–27

Ellingboe A.H. (1976): Genetics of host-parasite interactions. In: Heitefuss R., Williams P.H. (eds.): Encyclopedia of Plant Physiology, (NS), Physiological Plant Pathology. Springer Verlag, Heidelberg. 761–778

Ellis J., Jones D. (1998): Structure and function of proteins controlling strain-specific pathogen resistance. Curr. Opinion Plant Biol. **1**: 288–293

Ellis J.G., Lawrence G.J., Peacock W.J., Pryor A.J. (1988): Approaches to cloning plant genes conferring resistance to fungal pathogens. Annu.Rev.Phytopathol. **26**: 245–263

Godiard L., Grant M.R., Dietrich R.A., Kiedrowski S., Dangl J.L. (1994): Perception and response in plant disease resistance. Curr.Opinion Genet.Dev. **4**: 662–671

Hahlbrock K., Scheel D., Logemann E., Nürnberger T., Parniske M., Reinold S., Sacks W.R., Schmelzer E. (1995): Oligopeptide elicitor-mediated defense gene activation in cultured parsley cells. Proc.Natl.Acad.Sci.USA **92**: 4150–4157

Hammond-Kosack K.E., Jones J.D.G. (1996): Resistance Gene-Dependent Plant Defense Responses. Plant Cell **8**: 1773–1791

Hammond-Kosack K.E., Jones J.D.G. (1997): Plant resistance genes. Annu. Rev. Plant Physiol. Mol. Biol. **48**: 575–607

Holub E.B. (1997): Organization of resistance genes in *Arabidopsis*. In: Crute I.R., Holub E.B., Burdon J.J. (eds.): The Gene-for-Gene Relationship in Plant-Parasitic Interactions. CAB International, Oxon UK, New York NY. 5–26

Hulbert S., Pryor T., Hu G., Richter T., Drake J. (1997): Genetic fine structure of resistance loci. In: Crute I.R., Holub E.B., Burdon J.J. (eds.): The Gene-for-Gene Relationship in Plant-Parasitic Interactions. CAB International, Oxon UK, New York NY. 27–44

Hunt M.D., Neuenschwander U.H., Delaney T.P., Weymann K.B., Friedrich L.B., Lawton K.A., Steiner H.-J., Ryals J.A. (1996): Recent advances in systemic resistance research – a review. Gene **179**: 89–95

Innes R.W. (1998): Genetic dissection of R gene signal transduction pathways. Curr. Opinion Plant Biol. **1**: 299–304

Johnson R. (1992): Past, present and future opportunities in breeding for disease resistance, with examples from wheat. Euphytica **63**: 3–22

Karin M., Smeal T. (1992): Control of transcription factors by signal transduction pathways: the beginning of the end. Trends Biochem.Sci. **17**: 418–422

Keen N.T. (1997): Elicitor generation and receipt – the mail gets through, but how? In: Crute I.R., Holub E.B., Burdon J.J. (eds.): The Gene-for-Gene Relationship in Plant-Parasitic Interactions. CAB International, Oxon UK, New York NY. 379–388

Knogge W. (1991): Plant resistance genes for fungal pathogens – physiological models and identification in cereal crops. Z.Naturforsch. **46c**: 969–981

Kobe B., Deisenhofer J. (1994): The leucine-rich repeat: a versatile binding motif. Trends Biochem.Sci. **19**: 415–421

Kombrink E., Somssich I.E. (1995): Defense response of plants to pathogens. Adv.Bot.Res. **21**: 1–34

Lamb C.J. (1994): Plant disease resistance genes in signal perception and transduction. Cell **76**: 419–422

Lawrence G.J., Shepherd K.W., Mayo G.M.E., Islam M.R. (1994): Plant resistance to rusts and mildews: genetic control and possible machanisms. Trends Microbiol. **2**: 263–270

Lee H.I., Leon J., Raskin I. (1995): Biosynthesis and metabolism of salicylic acid. Proc. Natl.Acad.Sci.USA **92**: 4076–4079

Martin G., Frederick R., Tilmony R., Zhou J. (1996): Signal transduction events involved in bacterial speck disease resistance. In: Mills D., Kunoh H., Keen N.T., Mayama S. (eds.): Molecular Aspects of Pathogenicity and Resistance: Requirement for Signal Transduction. The American Phytopathological Society, St. Paul, Minnesota. 163–176

Moerschbacher B.M., Reisener H.-J. (1997): The hypersensitive resistance reaction. In: Hartleb H., Heitefuss R., Hoppe H.-H. (eds.): Resistance of Crop Plants against Fungi. Gustav Fischer, Jena, Stuttgart, Lübeck, Ulm. 126–158

Newton A.C. (1997): Cultivar mixtures in intensive agriculture. In: Crute I.R., Holub E.B., Burdon J.J. (eds.): The Gene-for-Gene Relationship in Plant-Parasitic Interactions. CAB International, Oxon UK, New York NY. 65–80

Pryor T. (1987): The origin and structure of fungal disease resistance genes in plants. Trends Genet. 3: 157–161

Pryor T., Ellis J. (1993): The genetic complexity of fungal resistance genes in plants. Advances in Plant Pathology 10: 281–305

Richter T.E., Ronald P.C. (2000): The evolution of disease resistance genes. Plant Mol.Biol. 42: 195–204

Ronald P.C. (1998): Resistance gene evolution. Curr. Opinion Plant Biol. 1: 294–298

Ryan C.A., Farmer E.E. (1991): Oligosaccharide signals in plants: a current assessment. Annu.Rev.Plant Physiol. 42: 651–674

Scheel D. (1998): Resistance response physiology and signal transduction. Curr. Opinion Plant Biol. 1: 305–310

Schulze-Lefert P., Peterhaensel C., Freialdenhoven A. (1997): Mutation analysis for the dissection of resistance. In: Crute I.R., Holub E.B., Burdon J.J. (eds.): The Gene-for-Gene Relationship in Plant-Parasitic Interactions. CAB International, Oxon UK, New York NY. 45–64

Shepherd K.W., Mayo G.M.E. (1972): Genes conferring specific plant disease resistance. Science 175: 375–380

Tenhaken R., Levine A., Brisson L.F., Dixon R.A., Lamb C. (1995): Function of the oxidative burst in hypersensitive disease resistance. Proc.Natl.Acad.Sci.USA 92: 4158–4163

Relevant papers

Apostol I., Heinstein P.F., Low P.S. (1989): Rapid stimulation of an oxidative burst during elicitation of cultured plant cells: role in defense and signal transduction. Plant Physiol. 90: 109–116

Bent A.F., Kunkel B.N., Dahlbeck D., Brown K.L., Schmidt R., Giraudat J., Leung J., Staskawicz B.J. (1994): *RPS2* of *Arabidopsis thaliana*: A leucine-rich repeat class of plant disease resistance genes. Science 265: 1856–1860

Bowling S.A., Clarke J.D., Liu Y., Klessig D.F., Dong X. (1997): The *cpr5* mutant of Arabidopsis expresses both NPR1-dependent and NPR1-independent resistance. Plant Cell 9: 1573–1584

Chandra S., Low P.S. (1995): Role of phosphorylation in elicitation of the oxidative burst in cultured soybean cells. Proc.Natl.Acad.Sci.USA 92: 4120–4123

Delaney T.P., Uknes S., Vernooij B., Friedrich L., Weymann K., Negrotto D., Gaffney T., Gutrella M., Kessmann H., Ward E., Ryals J. (1994): A central role of salicylic acid in plant disease resistance. Science 266: 1247–1250

de Nettancourt D. (1977): Incompatibility in Angiosperms. Springer-Verlag, Berlin, New York

Dickinson M.J., Jones D.A., JonesJDG (1993): Close linkage between the *Cf-2/Cf-5* and *Mi* resistance loci in tomato. Molec.Plant-Microbe Interact. 6: 341–347

Dietrich R.A., Delaney T.P., Uknes S.J., Ward E., Ryals J.A., Dangl J.L. (1994): *Arabidopsis* mutants simulating disease resistance response. Cell 77: 565–577

Dixon M.S., Jones D.A., Keddie J.S., Thomas.CM, Harrison K., Jones J.D.G. (1996): The tomato *Cf-2* disease resistance locus comprises two functional genes encoding leucine-rich repeat proteins. Cell 84: 451–459

Dong X., Mindrinos M., Davis K.R., Ausubel F.M. (1991): Induction of *Arabidopsis* defense genes by virulent and avirulent *Pseudomonas syringae* strains and by a cloned avirulence gene. Plant Cell 3: 61–72

Fisher R.A. (1961): A model for the generation of self sterility alleles. J.Theoret.Biol. 1: 411–414

Freialdenhoven A., Scherag B., Hollricher K., Collinge D.B., Thordal-Christensen H., Schulze-Lefert P. (1994): Nar-1 and Nar-2, two loci required for Mla(12)-specified race-specific resistance to powdery mildew in barley. Plant Cell 6: 983–994

Greenberg J.T., Ausubel F.M. (1993): Arabidopsis Mutants Compromised for the Control of Cellular Damage During Pathogenesis and Aging. Plant J. 4: 327–341

Greenberg J.T., Guo A., Klessig D.F., Ausubel F.M. (1994): Programmed death in plants: A pathogen triggered response activated coordinately with multiple defense functions. Cell 77: 551–563

Hammond-Kosack K.E., Jones D.A., Jones J.D.G. (1994): Identification of two genes required in tomato for full Cf-9-dependent resistance to Cladosporium fulvum. Plant Cell 6: 361–371

He S.Y., Huang H.-C., Collmer A. (1993): Pseudomonas syringae pv. syringae harpin$_{Pss}$: A protein that is secreted via the hrp pathway and elicits the hypersensitive response in plants. Cell 73: 1255–1266

Hockenbery D.M., Oltvai Z.N., Yin X.-M., Milliman C.L., Korsmeyer S.J. (1993): Bcl-2 functions in the antioxidant pathway to prevent apoptosis. Cell 75: 381–388

Horvitz H.R., Ellis H.M., Sternberg P.W. (1982): Programmed cell death in nematode development. Neurosci. Comment. 1: 867–869

Jabs T., Dietrich R.A., Dangl J.L. (1996): Initiation of runaway cell death in an Arabidopsis mutant by extracellular superoxide. Science 273: 1853–1856

Jörgensen J.H. (1988): Genetic analysis of barley mutants with modifications of powdery meldew resistance gene Ml-a12. Genome 30: 129–132

Jörgensen J.H. (1992): Discovery, characterization and exploitation of Mlo powdery mildew resistance in barley. Euphytica 63: 141–152

Jones D.A., Dickinson M.J., Balint-Kurti P.J., Dixion M.S., Jones J.D.G. (1993): Two complex loci revealed in tomato by classical and RFLP mapping of the Cf-2, Cf-4, Cf-5 and Cf-9 genes for resistance to Cladosporium fulvum. Molec.Plant-Microbe Interact. 6: 348–357

Jones D.A., Thomas C.M., Hammond-Kosak K.E., Balint-Kurti P.J., Jones J.D.G. (1994): Isolation of tomato Cf-9 gene for resistance to Cladosporium fulvum by transposon tagging. Science 266: 789–793

Mieth H., Speth V., Ebel J. (1986): Phytoalexin production by isolated soybean protoplasts. Z.Naturforsch.C 41c: 193–201

Mindrinos M., Katagiri F., Yu G.-L., Ausubel F.M. (1994): The A. thaliana disease resistance gene RPS2 encodes a protein containing a nucleotide-binding site and leucine-rich repeats. Cell 78: 1089–1099

Nasrallah J.B., Stein J.C., Kandasamy M.K., Nasrallah M.E. (1994): Signaling in the arrest of pollen tube development in self-incompatible plants. Science 266: 1505–1508

Nürnberger T., Nennstiel D., Hahlbrock K., Scheel D. (1995): Covalent cross-linking of the Phytophthora megasperma oligopeptide elicitor to its receptor in parsley membranes. Proc.Natl.Acad.Sci.USA 92: 2338–2342

Nürnberger T., Nennstiel D., Jabs T., Sacks W.R., Hahlbrock K., Scheel D. (1994): High affinity binding of a fungal oligopeptide elicitor to parsley plasma membranes triggers multiple defense responses. Cell 78: 449–460

Oppenheim R.W., Prevette D., Tytell M., Homma S. (1990): Naturally occuring and induced neuronal death in chick embryo in vivo requires protein and RNA synthesis: evidence for the role of cell death genes. Dev. Biol. 138: 104–113

Parker J.E., Hahlbrock K., Scheel D. (1988): Different cell-wall components from Phytophthora megasperma f.sp. glycinea elicit phytoalexin production in soybean and parsley. Planta 176: 75–82

Parker J.E., Schulte W., Hahlbrock K., Scheel D. (1991): An extracellular glycoprotein from Phytophthora megasperma f.sp. glycinea elicts phytoalexin synthesis in cultured parsley cells and protoplasts. Molec.Plant-Microbe Interact. 4: 19–27

Raff M.C. (1992): Social controls on cell survival and cell death. Nature 356: 397–400

Raff M.C., Barres B.A., Burne J.F., Coles H.S., Ishizaki Y., Jacobson M.D. (1993): Programmed cell death and the control of cell survival: Lessons from the nervous system. Science **262**: 695–700

Rohwer F., Fritzemeier K.-H., Scheel D., Hahlbrock K. (1987): Biochemical reactions of different tissues of potato (*Solanum tuberosum*) to zoospores or elicitors from *Phytophthora infestans*. Accumulation of sesquiterpenoid phytoalexins. Planta **170**: 556–561

Rommens C.M.T., Rudenko G.N., Djikwel P.P., van Haaren M.J.J., Ouwerkerk P.B.F., Block K.M., Nijkamp H.J.J., Hille J. (1992): Characterization of the Ac/DS behaviour in transgenic tomato plants, using plasmid rescue. Plant Mol.Biol. **20**: 61–70

Salmeron J.M., Barker S.J., Carland F.M., Mehta A.Y., Staskawicz B.J. (1994): Tomato mutants altered in bacterial disease resistance provide evidence for a new locus controlling pathogen recognition. Plant Cell **6**: 511–520

Scheel D., Colling C., Hedrich R., Kawalleck P., Parker J.E., Sacks W.R., Somssich I.E., Hahlbrock K. (1991): Signals in plant defense gene activation. In: Hennecke H., Verma D.P.S. (eds.): Advances in Molecular Genetics of Plant-Microbe Interactions. Dordrecht, Kluwer Academic Publishers, Netherlands. 373–380

Sharp J.K., Valent B., Albersheim P. (1984): Purification and partial characterization of a β-glucan fragment that elicits phytoalexin accumulation in soybean. J.Biol.Chem. **259**: 11312–11320

Sutherland M.W. (1991): The generation of oxygen radicals during host responses to infection. Physiol.Mol.Plant Path. **39**: 79–94

Torp J., Jörgensen J.U. (1986): Modification of barley powdery mildew resistance gene Ml-a12 by induced mutation. Can.J.Genet.Cytol. **28**: 725–731

Van den Beek J.G., Verkerk R., Zabel P., Lindhout P. (1992): Mapping strategy for resistance genes in tomato based on RFLPs between cultivars: resistance to *Cladosporium fulvum* on chromosome 1. Theor.Appl.Genet. **84**: 106–112

Veis D.J., Sorensen C.M., Shuter J.R., Korsmeyer S.J. (1993): *Bcl-2*-deficient mice demonstrate fulminant lymphoid apoptosis, polycystic kidneys, and hypopigmented hair. Cell **75**: 229–240

Xing T., Higgins V.J., Blumwald E. (1997): Race-specific elicitors of *Cladosporium fulvum* promote translocation of cytostolic components of NADPH oxidase to the plasma membrane of tomato cells. Plant Cell **9**: 249–259

Yu I.-c., Parker J., Bent A.W. (1998): Gene-for-Gene disease resistance without the hypersensitive response in *Arabidopsis dnd1* mutant. Proc.Natl.Acad.Sci.USA **95**: 7819–7824

Yu I.-c., Fengler K.A., Clough S.J., Bent A.W. (2000): Identification of Arabidopsis mutants exhibiting an altered hypersensitive response in gene-for-gene disease resistance. Molec.Plant-Microbe Interact. **13**: 277–286

9.1.5
Coevolution of Resistant Host Plants and Their Specific Virulent Pathogens in Agriculture

Intensive agricultural production frequently makes use of genetically uniform cultivars of high performance grown as single crops over large areas. In many cases the cultivars carry a gene for race-specific resistance to protect the crop against a particular pathogen. However, after a few years of successful harvests, the cultivar may lose its resistance. To maintain crop yield, the farmer switches to a cultivar with another gene for race-specific resistance. However, this new cultivar may in turn become susceptible after several years of cultivation and the farmer is confronted with the same problem. This unwelcome but frequently observed loss of race-specific resistance is of course due to the occurrence in the pathogen population of spontaneous mutations to specific virulence (see also Sect. 9.1.1). Monocultures of race-specific resistant cultivars very efficiently select virulent pathogen races able to negate the race-specific resistance gene in their host plant.

The continued introduction of newly bred, race-specific resistant cultivars is responsible for a man-made coevolution of race-specific resistant cultivars and spontaneously emerging, specific virulent pathogens. The driving forces are the high rates of reproduction and mutation to specific virulence in pathogens and the continuous introduction in agriculture of new race-specific resistant cultivars.

Nearly all individuals in a population of phytopathogenic fungi share the same set of avirulence genes, the products of which may act as avirulence factors. A pathogen is avirulent because it interacts with a race-specific resistant host plant carrying a particular receptor able to recognize one of the pathogen's avirulence factors. The avirulence factor is a "label" by which the resistant plant recognizes the attacking homologous pathogen. If we suppose that avirulence genes of pathogens are wild type genes present in all phytopathogenic fungi then, in contrast, the race-specific resistance genes of plants came into being by mutation. Race-specific resistant cultivars were selected by plant breeders hoping to obtain cultivars with genetically stable resistance. However, this was not to be.

The development of host-pathogen coevolution can be divided into a series of steps (*Fig. 11*). For example the starting point might be a wheat cultivar of genotype $r_1r_2r_n$ completely susceptible to a rust pathogen. To control the rust disease plant breeders screen in a **first step** other material to find a gene for resistance, R_1, which is introduced by hybridization. The gene R_1 produces a resistance factor Rcpt1 by which the resistant cultivar is able to recognize the attacking rust's avirulence factor A_1 the product of its avirulence gene Avr_1. In this way defense reactions in the wheat plant are released against the attacking pathogen. The **second step** of coevolution is the selection of spontaneous pathogen mutants with specific virulence of genotype $avr_1Avr_2Avr_n$. These mutants have a defective avirulence gene avr_1 which is unable to synthesize an active avirulence product A_1. The mutated pathogen that synthesizes only the defective product a_1 escapes recognition by the resistance factor of the R_1 resistant host plant, and parasitizes it (see also Sect. 9.1.1).

The formerly resistant wheat cultivar suffers a loss in grain yield, and so a new cultivar of genotype R_1R_2 with an additional gene, R_2, for race-specific resistance, is selected and released. This is the **third step** in coevolution and is similar to the first step when the gene R_1 was introduced. The doubly race-specific resistant cultivar of genotype R_1R_2 recognizes and defends itself against pathogens of genotype $avr_1Avr_2Avr_n$ (*Fig. 11*), because the Rcpt2 receptor recognizes the avirulence product A_2 determined by gene Avr_2. The specific virulence allele avr_1 has no effect since avirulence is epistatic to virulence. The **fourth step** in coevolution corresponds to the second step: R_1R_2 cultivars ultimately select specific virulent pathogens of genoptype $avr_1avr_2Avr_n$ (*Fig. 11*), unable to synthesize active A_1 and A_2 products thereby preventing their recognition by resistance factors $Rcpt_1$ and $Rcpt_2$. The next step, introduction of a third race-specific resistance gene R_n, not shown in *Fig. 11*, would then be followed by selection of a pathogen with specific avirulence alleles $avr_1/avr_2/avr_n$.

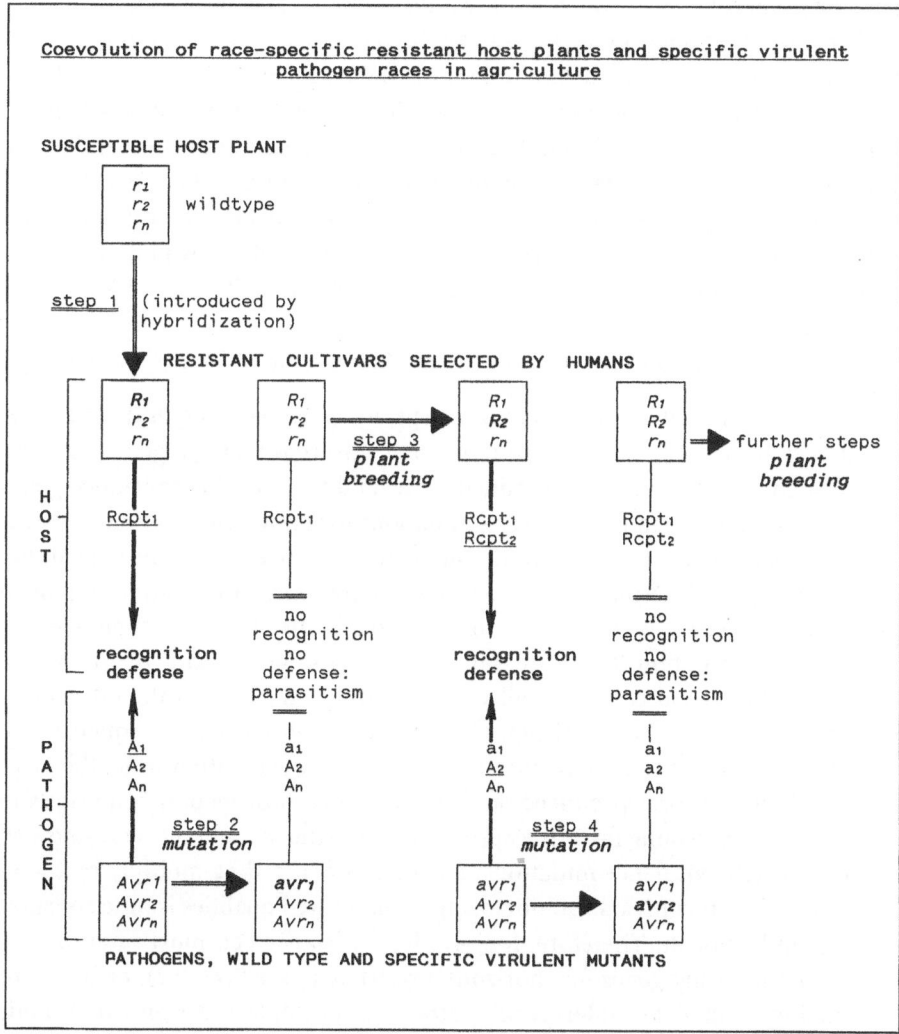

Figure 11. Coevolution in agriculture of race-specific resistant host plant and homologous pathogen species. Coevolution begins with infection of a susceptible cultivar carrying no race-specific resistance genes (r_1, r_2, r_n). Plant breeders introduce by hybridization the race-specific resistance gene R_1 to defend particular pathogen species. However, the respective pathogen population carrying avirulence genes Avr_1, Avr_2 or Avr_n may contain rare specific virulent mutants avr_1 and avr_2. These produce only defective products a_1 and a_2, or none at all. Since they are not recognized by the receptors produced by the resistance genes R_1 or R_2 the pathogen mutants are selected and will multiply on these hosts. Gene products recognizing each other are underlined. For more details, see text.

Coevolution of pathogens and their host plants will probably continue for as long as man breeds new race-specific resistant cultivars for use in agricultural production. Long-lasting resistance cannot be achieved by breeding for race-specific resistance. Breeding novel genotypes based on race-specific resistance in turn generates novel types of specific virulent mutant pathogens and new specific virulence alleles. This type of evolution is unlikely to occur in a natural ecosystem, since the host plant population would rarely be genetically uniform.

One important question remains to be discussed. Pathogens carrying the wild type gene Avr_1 are unable to multiply on R_1 race-specific resistant host plants. Given this condition, how did an avr_1 specific virulent mutant arise when the wild type Avr_1 from which it is derived is excluded from reproduction and hence from the possibility of mutating? There are two explanations of how specific virulent mutants may appear on race-specific resistant cultivars. The mutant could have been carried in from a crop growing elsewhere or the new pathotype may have arisen by mutation if the race-specific resistance permitted low-level pathogen growth and limited spore production.

Farmers have tried to prolong the life of race-specific resistance in two ways:

(1) Two or more different genes for race-specific resistance were introduced together in one cultivar and were not first released in single gene cultivars. Overcoming resistance determined by several race-specific resistance genes requires that the pathogen carries mutations to specific virulence in all avirulence genes corresponding to the cultivar's resistance genes. The probability of acquiring all of these specific virulent mutations, either in one step or in successive steps, is the product of the probabilities for each single specific virulence mutation. The more different genes for race-specific resistance are bred into one cultivar, also called "pyramiding", the smaller the probability of a virulent pathogen arising. However, the high mutation frequencies of most pathogens to specific virulence, their short generation time, the large numbers of spores produced, and the favorable propagation conditions in single-crop farming favor the appearance of pathogen mutants carrying several specific virulence mutations. Furthermore, breeding multiply race-specific resistant cultivars without losing other desired qualities is time consuming and difficult. Therefore, long-lasting resistance can most often be obtained by using genes for horizontal resistance (see Sect. 9.2), or by introducing, either by interspecific crossing, protoplast fusion or genetic transformation, resistance determinants from other plant species that control the mechanisms of basic resistance.

(2) The farmer grows a mixture of different cultivars each of which carries a different race specific resistance gene. Needless to say these cultivars should be of similar quality and maturity so that they can be harvested and processed together. A similar approach is provided by a so-called multiline variety, made up of components which are isogenic except that they carry dif-

ferent race specific resistance genes. If the pathogen population contains pathotypes able to colonize only one component of the mixture or multiline, there is a decreased chance of their proliferation because although they can sporulate and reproduce on one component they cannot do so on the other components with different race-specific resistance. Furthermore, when replanting in the following season the farmer can replace individual components which were susceptible. Cultivar mixtures and multilines were also used to reduce application of fungicides. The efficacy of cultivar mixtures and multilines in reducing disease depends in large part on the absence of multiply virulent pathotypes able to colonize most or all of the components. However, like multiply resistant varieties, cultivar mixtures also select pathotypes that have accumulated specific virulence genes. There may also be economic risks, in growing mixtures and multilines if the harvested end product cannot meet a required quality standard.

Plant species in natural plant populations are genetically heterogeneous, quite unlike crop monocultures. Although selection will occur in natural plant populations for various pathotypes they do not present the opportunity for unhindered multiplication provided by race-specific resistant monocultures.

The genetic composition of a pathogen population will be determined in large part by the genetic composition of the host plant population. The dynamics of population structure are complex and take account of findings from population genetics, epidemiology, and ecology. Predicted alterations in pathogen and plant populations obtained from mathematical models can be compared with the results obtained in field experiments. The aim of such studies is to provide a reliable basis for forecasting disease development and designing better methods of crop protection. Although this topic is beyond the scope of this book some relevant reviews are given at the chapter's end.

We believe that it is important to distinguish between race-specific resistance genes and other genetic determinants in host plants sometimes called "resistance genes." The latter determinants are neither single genes nor are they related to race-specific resistance. In many cases they were introduced by interspecific hybridization or chromosome engineering, techniques used to add or substitute chromosomal fragments derived from wild plant species that are infection-resistant. As a rule, in such cases not a single resistance gene but large DNA fragments containing several genes expressing horizontal resistance or race-non-specific resistance were transferred (see Sect. 9.2). In other cases, mechanisms for basic resistance not present in the recipient plant were transferred. These mechanisms include the ability to produce host-specific effectors such as particular phytoalexins, or the deposition of suberin, lignin, or phenols in cell walls. In contrast to race-specific resistance, they are very stable genetic traits, since they cannot be negated by a single mutation in the (haploid) pathogen. Rather, overcoming such basic resistance would require the evolution of quite new pathogenicity genes. A new mechanism for basic resistance should be effective not only against one particular *forma specialis* of a pathogen, as in non-race-specific or horizontal resistance, but also against many different pathogen species. However, breeding such new resistant cultivars may have serious drawbacks, including the loss of desired quality and other characteristics.

Reviews

Bayles R.A., Clarkson J.D.S., Slater S.E. (1997): The UK cereal pathogen virulence survey. In: Crute I.R., Holub E.B., Burdon J.J. (eds.): The Gene-for-Gene Relationship in Plant-Parasitic Interactions. CAB International, Oxon UK, New York NY. 103–117

Brown J.K.M., Foster E.M., O'Hara R.B. (1997): Adaption of powdery mildew populations to cereal varieties in relation to durable and non-durable resistance. In: Crute I.R., Holub E.B., Burdon J.J. (eds.): The Gene-for-Gene Relationship in Plant-Parasitic Interactions. CAB International, Oxon UK, New York NY. 119–138

Browning J.A., Frey K.J. (1969): Multiline cultivars as a means of disease control. Annu. Rev.Phytopathol. 7: 355–382

Fischbeck G. (1997): Gene management. In: Hartleb H., Heitefuss R., Hoppe H.-H. (eds.): Resistance of Crop Plants against Fungi. Gustav Fischer, Jena, Stuttgart, Lübeck, Ulm. 349–377

Frank S.A. (1992): Models of plant-pathogen coevolution. Trends Genet. 8: 213–219

Hartleb H., Heitefuss R. (1997): Abiotic and biotic influences on resistance of crop plants against fungal pathogens. In: Hartleb H., Heitefuss R., Hoppe H.-H. (eds.): Resistance of Crop Plants against Fungi. Gustav Fischer, Jena, Stuttgart, Lübeck, Ulm. 298–326

Heath M.C. (1991): Evolution of resistance to fungal parasitism in natural ecosystems. New Phytol. 119: 331–343

Hovmoller M.S., Ostergard H., Munk L. (1997): Modelling virulence dynamics of airborne plant pathogens in relation to selection by host resistance in agricultural crops. In: Crute I.R., Holub E.B., Burdon J.J. (eds.): The Gene-for-Gene Relationship in Plant-Parasitic Interactions. CAB International, Oxon UK, New York NY. 173–190

Johnson T. (1961): Man-guided evolution in plant rusts. Science 133: 357–362

Kolmer J.A. (1997): Virulence dynamics and genetics of cereal rust populations in North America. In: Crute I.R., Holub E.B., Burdon J.J. (eds.): The Gene-for-Gene Relationship in Plant-Parasitic Interactions. CAB International, Oxon UK, New York NY. 139–156

Limpert B., Bartos B. (1997): Analysis of pathogen virulence as decision support for breeding and cultivar choice. In: Hartleb H., Heitefuss R., Hoppe H.-H. (eds.): Resistance of Crop Plants against Fungi. Gustav Fischer, Jena, Stuttgart, Lübeck, Ulm. 401–424

Newton A.C. (1997): Cultivar mixtures in intensive agriculture. In: Crute I.R., Holub E.B., Burdon J.J. (eds.): The Gene-for-Gene Relationship in Plant-Parasitic Interactions. CAB International, Oxon UK, New York NY. 65–80

Parlevliet J.E. (1997): Durable resistance. In: Hartleb H., Heitefuss R., Hoppe H.-H. (eds.): Resistance of Crop Plants against Fungi. Gustav Fischer, Jena, Stuttgart, Lübeck, Ulm. 238–253

Thompson J.N., Burdon J.J. (1992): Gene-for-gene coevolution between plants and parasites. Nature 360: 121–125

Welz H.G., Kranz J. (1997): How resistance affects disease epidemics in crops. In: Hartleb H., Heitefuss R., Hoppe H.-H. (eds.): Resistance of Crop Plants against Fungi. Gustav Fischer, Jena, Stuttgart, Lübeck, Ulm. 327–348

Wolfe M.S., Finckh M.R. (1997): Diversity of host resistance within the crop: effects on host, pathogen and disease. In: Hartleb H., Heitefuss R., Hoppe H.-H. (eds.): Resistance of Crop Plants against Fungi. Gustav Fischer, Jena, Stuttgart, Lübeck, Ulm. 378–400

9.1.6
Preservation of *Avr* Genes in Pathogen Populations and Their Possible Functions

A plant's resistance to a homologous pathogen is only brought into play after it has recognized one of the pathogen's avirulence products. However, recognition depends on the presence of a corresponding, race-specific resistance gene in the plant. The avirulence product serves as a signal or "label" indicating an attacking pathogen (see also Chap. 9, Sect. 9.1.1). As discussed already in Sects. 9.1.1 and 9.1.5, each pathogen has many different genes able to function as avirulence genes and each produces a specific avirulence factor. Furthermore, there are many different race-specific resistance genes each of which corresponds specifically to a particular avirulence gene and its avirulence product. As a consequence each pathogen avirulence gene may also function as a lethal factor when the pathogen infects a host plant with the corresponding race-specific resistance gene. This raises two questions: (1) what is the nature of avirulence genes and (2) why are they still present in pathogens even though they are lethal when infecting certain host plants? One might expect that during evolution there would have been selection against *Avr* genes because of recognition and its lethal effects. However, such counterselection was obviously incomplete, since phytopathogenic fungi still possess many avirulence genes producing avirulence factors. Do these avirulence factors offer some selective advantage to pathogens infecting susceptible host plants, or does their absence impose a selective disadvantage?

Avirulence products were thought to be products of secondary metabolism. For example the *Vb/Pc-2* locus of the oat cultivar Victoria recognizes two quite different products: the avirulence factor of *Puccinia coronata* and the host-selective toxin victorin C, produced by *Cochliobolus heterostrophus* (Helminthosporium victoriae) (see also Chap. 8, Sect. 8.2 and Chap. 9, Sect. 9.1.3.3). Since victorin C is a product of secondary metabolism, it was thought that the avirulence product of *P. coronata*, recognized by the same chromosomal locus, might have a similar derivation. However, products of secondary metabolism, e.g., antibiotics, toxins, or substances with similar antagonistic effects, are produced via pathways made up of many synthetic steps controlled by different genes. If only a single gene encodes an avirulence factor, as proposed by the gene-for-gene hypothesis, the primary product of an avirulence gene should be an RNA or a polypeptide or protein, not a product of secondary metabolism synthesized by a pathway controlled by several genes. Indeed, the few avirulence products so far identified are polypeptides encoded by single genes (see also Sect. 9.1.3.1). Of course this may be the result of chance considering the small number of avirulence products of phytopathogenic fungi that have been identified to date. For this reason the question whether substances other than primary gene products could act as avirulence factors is still relevant. To pose the question another way:

Are there other, possibly more complex, connections which only formally entail a gene-for-gene relationship? (See also Sect. 9.1.1, Person's matrix.)

There are some striking exceptions to the rule that only one gene participates in the determination of avirulence. Crossing experiments showed that the expression of avirulence may be suppressed by another gene present in the pathogen. This can be explained if the avirulence product is synthesized via a metabolic pathway, i.e., the final host-pathogen interaction proceeds as a "gene-for-*gene-pathway*" (see Sect. 9.1.8) instead of simply "gene-for-gene". Recent refinements in selection methods and the application of molecular genetic techniques, have revealed that some chromosomal loci encoding the synthesis of secondary metabolites, such as phytotoxins, consist of a cluster of several closely linked genes. Accordingly, some interactions releasing race-specific resistance considered until now as gene-for-gene interactions may be dependent on a gene-for-gene-pathway.

Why pathogens still harbor avirulence genes has been widely discussed and investigated experimentally. For example *Avr* genes might furnish pathogens with some selection advantage or fulfill functions distinct from triggering defense reactions in race-specific resistant plants. It was supposed that pathogens carrying wild type *Avr* genes would exhibit greater parasitic fitness when colonizing susceptible host plants compared with pathogens carrying defective, or specific virulent, alleles. The answer to this question is still somewhat controversial, partially depending on the experimental system employed for testing.

During axenic cultivation of pathogens carrying either *Avr* wild type or *avr* mutant alleles showed no differences in growth rates. Also the phenotypes of both *Avr* or *avr* alleles harboring pathogens were indistinguishable. Thus, the *Avr* gene and its ensuing function seemed to be dispensable. However, differences in growth rates of less than 1 % can hardly be recognized within short-term experiments. Furthermore, even much smaller differences in growth rates effective may lead, over long periods, to selection of pathogens carrying a particular allele, e.g., a wild type gene *Avr*. Such selection processes may establish a particular population structure or even promote evolution of new pathogen genotypes. – In summary, short-term experiments measuring growth rates in axenic culture are unfit for detecting differences in parasitic fitness.

Another type of experiment proved much more sensitive, namely competition between *Avr* and *avr* variants of a pathogen infecting susceptible and resistant host plants when afterwards the numbers of vegetative spores produced *in planta* of *Avr* and avr genotype were screened. Such competition experiments were carried out with pigment mutants of *Cladosporium fulvum* inoculated to the cotyledons of aseptically grown resistant or susceptible tomato seedlings (Day 1968). Equal numbers of conidia of the wild type and a non-pigmented mutant were placed on seedlings grown in the light in large diameter tubes. The proportions of the two components in each mixture were estimated by harvesting the conidia produced after 21 days, plating them on medium, and counting the numbers of colonies of each color after incubation. Non-pigmented mutants were less vigorous than wild type

C. fulvum carrying identical *Avr* alleles. However, the high parasitic fitness of the wild type pigmented *Avr* strain was diminished to a level below that of the non-pigmented *Avr* mutant when the former carried an *avr* allele.

Similar investigations demonstrated that certain genes for avirulence, that are strongly selected against employing cultivars with corresponding race-specific resistance genes, reappear in a pathogen population as soon as the *R* gene selection pressure is removed. This may happen by infecting cultivars harboring no corresponding *R* genes, or by infecting multiline cultivars or cultivar mixtures (see Chap. 9, Sect. 9.1.5). Such changes in the genetic structure of a pathogen population were observed with *Phytophthora infestans*, *Puccinia graminis* f.sp. *graminis*, and *P. graminis* f.sp. *avenae*, and were reviewed by Vanderplank 1968, Browning and Frey 1969, Watson 1970, and Gabriel 1989.

The reappearance of previously counterselected *Avr* alleles has been called "stabilizing selection" by Vanderplank. However, it has also been shown that a defective *avr* allele is not always responsible for reduced parasitic fitness. Crosses among races of *Erysiphe graminis* f.sp. *tritici* carrying specific virulent mutations revealed a trait for reduced parasitic fitness separable from that for specific virulence (Bronson and Ellingboe 1986). Therefore, the hypothesis that stabilizing selection depends only on specific virulence alleles, reducing parasitic fitness of a pathogen, is still controversial.

However, it was recently shown that a gene product of a necrotrophic pathogen, the phytotoxic protein NIP1 of *Rhynchosporium secalis*, recognized in a race-specific resistant plant as an avirulence factor may also function as a virulence factor. When infecting the susceptible host plant it intensifies the expression of disease symptoms and evidently improves the fitness of the parasite (Rohe et al. 1995; see also Sect. 9.1.3.1).

Maintenance of *Avr* genes in a reproducing pathogen population should depend on the fraction of host plants containing the corresponding *R* genes, since only this fraction is counterselective. Thus it is to be expected that the ratio of Avr_n to avr_n alleles in the pathogen population will be correlated with the ratio of the corresponding alleles R_n and r_n in the host population. A stable self-adjusting equilibrium between specific virulent pathogens and race-specific resistant host plants should occur in natural ecosystems.

Factors likely to contribute to the maintenance of a stable equilibrium between the different genotypes of pathogens and their host plants include the genetic polymorphism of the gene-for-gene system, such as well the relative frequencies of multiple allelic avirulence and resistance loci in the pathogens and host plants; their different mutation frequencies; and differences in parasitic fitness between *Avr* and *avr* pathogen genotypes. Quantitative evaluations of the interactions between all of these parameters have been attempted employing mathematical models for predicting the coevolution of pathogens and their host plants in natural ecosystems. Thus, one mathematically formulated hypothesis states that, because of the polymorphic nature of host-pathogen interaction and the differences in mutation rates among the participating genes, minor differences in the fitness of specific virulent pathogens, together with small disadvantages in the growth of plants carrying resistance genes, may result in the maintenance of avirulence alleles within a natural ecosystem.

The unusually high mutation rate of avirulence genes (see also Sect. 9.1.4) should be noted in this context. In *Puccinia graminis*, for example, the mutation rate is about 10^{-5}, a value 10 times higher than in other genes. In other host-pathogen systems mutation frequencies may even reach 10^{-3} to 10^{-2}. That a pathogen can tolerate such a high mutation rate in some of its genes is only conceivable if it is vital for survival. Thus although *avr* alleles may lower fitness on susceptible plants *Avr* alleles are lethal on race-specific resistant plants. Under these circumstances the high mutability of avirulence genes is advantageous for pathogen survival: the fraction of pathogens that carry defective or specific virulent *avr* alleles safeguard pathogen survival.

The high mutation rate of avirulence genes suggests that special mechanisms are responsible for it. For example, in the phytopathogenic bacterium *Xanthomonas campestris* pv. *vesicatoria*, (*Xcv*) causing bacterial spot disease on pepper, *Capsicum annum*, transposon insertion into the avirulence gene *AvrBs1* has been observed to occur at different sites and at a frequency about ten times higher than in backgrounds lacking the transposon (Kearney et al. 1988). The transposon identified was the insertion sequence IS476, 1,2 kb in size, which inserted in each mutant at different sites within the *AvrBs1* avirulence gene as proven by restriction digestes and Southern blotting of the *Xcv* DNA. However, this is a very special case not suitable for generalization because it turned out that the transposon was introduced into *C. campestris* via the transmissible plasmid determining copper resistance which carries three IS476 elements. Copper was used for years in Florida as a chemical control for bacterial spot disease and *Xcv* was under selective pressure for copper resistance. Correspondingly, all copper resistant *Xcv* strains were shown to contain IS476 suggesting that it was spread by the copper-selected transmission of plasmid pXvCu1.

The DNA of avirulence genes in phytopathogenic fungi is thought to harbor highly variable regions, giving rise to the high mutation rates from *Avr* to *avr*. However, the nature of these regions is still under discussion (illegitimate recombination, see Sect. 9.1.4). Their existence may be the result of a selection process guaranteeing pathogen survival (see above).

Reviews

Atkinson M.M. (1993): Molecular mechanisms of pathogen recognition by plants. Advances in Plant Pathology 10: 35–64

Browning J.A., Frey K.J. (1969): Multiline cultivars as a means of disease control. Annu. Rev.Phytopathol. 7: 355–382

Crute I.R. (1985): The genetic bases of relationships between microbial parasites and their hosts. In: Fraser R.S.S. (ed.): Mechanisms of Resistance to Plant Diseases. Martinus Nijhoff/Dr.W.Junk, Dordrecht, Boston, Lancaster. 81–142

De Wit P.J.G.M. (1992): Molecular characterization of gene-for-gene systems in plant-fungus interactions and the application of avirulence genes in control of plant pathogens.
Annu.Rev.Phytopathol. **30**: 391-418

Frank S.A. (1992): Models of plant-pathogen coevolution. Trends Genet. **8**: 213–219

Gabriel D.W. (1989): Genetics of plant parasite populations and host-parasite specificity.
In: Kosugue T., Nester E.W. (eds.): Plant-Microbe Interactions: Molecular and Genetic
Perspectives. Macmillan, New York. 343–379

Gabriel D.W., Rolfe B.G. (1990): Working models of specific recognition in plant-microbe
interactions. Annu.Rev.Phytopathol. **28**: 365–391

Stone M.J., Williams D.H. (1992): On the evolution of functional secondary metabolites
(natural products). Mol.Microbiol. **6**: 29–34

Vanderplank J.E. (1963): Plant Diseases: Epidemics and Control. Academic Press, New
York, London

Vanderplank J.E. (1968): Disease Resistance in Plants. Academic Press, New York, London

Watson I.A. (1970): Changes in virulence and population shifts in plant pathogenesis. An-
nu.Rev.Phytopathol. **8**: 209–230

Watson I.A., Luig N.H. (1968): The ecology and genetics of host pathogen relationships in
wheat rusts in Australia. In: Finlay K.W., Shepherd K.W. (eds.): Pro. Int. Wheat Genet.
Symp., 3rd,. Aust. Acad. Sci., Canberra. 479 pp.

Relevant papers

Bronson C.R., Ellingboe A.H. (1986): The influence of four unnecessary genes for virulence
on the fitness of *Erysiphe graminis* f. sp. *tritici*. Phytopathology **76**: 154–158

Day P.R. (1968): Plant disease resistance. Sci. Progr., Oxf. **65**: 357–370

Kearney B., Ronald P.C., Dahlbeck D., Saskawicz B.J. (1988): Molecular basis for evasion of
plant host defence in bacterial spot disease of pepper. Nature **332**: 541–543

Rohe M., Gierlich A., Hermann H., Hahn M., Schmidt B., Rosahl S., Knogge W. (1995): The
race-specific elicitor, NIP1, from the barley pathogen, *Rhynchosporium secalis*, determines avirulence on host plants of the *Rrs1* resistance genotype. EMBO J. **14**: 4168–4177

9.1.7
The Evolution of Race-Specific Resistance in Natural Ecosystems

In natural ecosystems pathogens normally do much less damage than in single
crop farming of genetically uniform, race-specific resistant cultivars. The occurrence and relevance of race-specific resistance in natural ecosystems has still only
been explored to a limited extent. The most thoroughly analyzed system to date is
that of groundsel, *Senecio vulgaris,* infected with the powdery mildew, *Erysiphe
fisheri*. However, all results so far obtained indicate that this resistance type is
present in all wild plant populations with considerable genetic polymorphism.
This is of course not surprising considering the success that modern plant breeders have had in locating race specific resistance in collections of wild relatives of
crop plants.

The persistence of plants with race-specific resistance mutations will mainly
be influenced by the genetic polymorphism associated with this resistance type
and the balance between the emergence of novel resistance specificities through
mutation or unequal crossing over and the occurrence of specific virulent pathogens arising either by mutation or by immigration from elsewhere. These factors

should lead to an equilibrium between host plants and pathogens which depends on the numbers of alleles for susceptibility and resistance in plants and for avirulence or specific virulence in pathogens. The equilibrium will also depend on the advantage or disadvantage, i.e. the "cost to fitness," conferred by the new host plant and/or pathogen genotypes. A stable equilibrium should guarantee survival of both host plants and their pathogens. Indeed, the few investigations undertaken so far in wild type populations confirm the existence of such equilibria between plants and their pathogens.

Additional genetic traits that play a role include growth rate, generation time, and numbers of seeds or spores produced and their viability. External factors such as climate, nutrition and competition with other plants or pathogens will also influence the equilibrium.

A host-pathogen system susceptible to internal and external factors has only a limited buffering capacity when confronted with dramatic alterations of these factors. This means that equilibrium can only be restored after modest alterations. After severe disturbances the equilibrium may break down completely and the pathogen may overwhelm its host compromising its own survival in the process. Complete breakdown may lead to extinction. However, a new stable ecosystem could arise from a few surviving host plants and some different, specific virulent pathogens immigrating from elsewhere.

Building up a novel ecosystem under such conditions would mean starting with a restricted spectrum of alleles for resistance, specific virulence, and other characteristics. The smaller the number of genotypes surviving the break down of the previous host-pathogen system the fewer pathogen genotypes will immigrate from outside and become established. A novel host-pathogen system that develops by mutation and selection, once it reaches equilibrium, may exhibit a spectrum of resistance and virulence alleles that are quite different from the prior system.

Why do we rarely find wild plant populations with only a single type of race-specific resistance? Why is genetic polymorphism for race-specific resistance characteristic of wild populations? A possible answer may come from investigations of the evolution of gene-for-gene interactions in natural pathosystems (Burdon et al. 1996): Wild populations of plants and pathogens do not present large coherent areas with uniformly distributed frequencies of alleles for resistance and specific virulence. Rather, they consist of a mosaic of small subpopulations, each with qualitatively, as well as quantitatively, different allele frequencies for resistance and specific virulence. They make up a large, so-called metapopulation. The different small component subpopulations, or demes, often overlap. If the plant-pathogen equilibrium breaks down it usually does so only in one of the demes, in which a new equilibrium will subsequently be established by immigration of pathogens, and no doubt also of host genes through outcrossing, from neighboring demes. The newly established deme may then contain quite different allele frequencies. The metapopulation structure, with

its many different subpopulations, encompasses many different race-specific resistance types. However, each deme is unique. For example, if in one deme a
virulent pathogen mutant causes a local population breakdown, this will extin-
guish the corresponding race-specific resistant plants only in that confined area,
but is unlikely to lower the proportion of these plants within the metapopulation.
After the local breakdown of the subpopulation, or deme, a different and newly
structured host-pathogen equilibrium will slowly develop again in this confined
area.

In rare cases even wild populations of plants may suffer serious damage similar to that observed for homogeneous, race-specific cultivars attacked by a
specific virulent pathogen race. Serious damage occurs although the wild plant
populations are not homogeneously race-specific resistant and the attacking
pathogens not race-specific virulent. A striking example of such serious damage
of a wild type plant population has been reported by Weste and Kennedy (1997)
from Australia after devastation of native trees and shrubs of different genera by
Phytophthora cinnamomi in the Grampians, Victoria. After complete devastation
of the vegetation a 100 % contamination of the soil with *P. cinnamomi* was found
during 1976/77 to 1984. However, in 1995 *P. cinnamomi* was found in only 15,6 %
of 345 of soil and root samples and was present in 28,6 % of infected quadrats
tested as compared with 100 % ten to twenty years before. The present structure
and composition of plant species was compared between the still infected and the
uninfected plots. The infected quadrats contained several dead trees and trees
with reduced crown density and dieback of the major branches (*Eucalyptus bax-
teri* and *E. obliqua*) and 8 years old saplings and seedlings showed few or no
infection symptoms despite isolation of *P. cinnamomi* from their roots. The understorey of infected plots comprised a dense growth of field-resistant plants
such as *Leptospermum* spp. and sedges entwined with dodder laurel with
only 4 % of susceptible plant species, whereas the uninfected plots contained
a species-rich heath flora of which 54 % were known to be susceptible to *P. cinnamomi*. Regeneration and continuing survival of 24 susceptible species from 11
different families were recorded as small plants on various infected quadrats.
However, 12 susceptible species from 7 different families failed to regenerate.
This suggests that pathogen populations increased with the renewed supply
of susceptible roots, but independent of pathogen potential, the emerging susceptible species could not compete effectively against the dense field-resistant
understorey of the previously infected plots in order to re-establish the original
diverse heathland understorey.

The present status and the alteration undergone since suggested the following
conclusions: As far as it concerns the pathogen a decline in population density
and in inoculum levels occurred. Also a decline in disease has happened, associated with the reduction in susceptible host roots, encouraging selection of saprophytic vigour in the hyphae since *P. cinnamomi* was isolated from the roots of
susceptible plants showing no disease symptoms. Also continuous or variable

reactions to the soil microbial population outcompeting, antagonizing or even parasitizing the hyphae of *P. cinnamomi* may have occured; or increased resistance among the progeny of some susceptible plants which enables their regeneration and survival in the presence of the pathogen. – The data indicate that factors controlling the pathogen, the environment, the availability of susceptible host roots, host plant resistance, and more probably a combination of these have changed during the time interval of 10 to 20 years since the first observations were recorded. The question arises as to which of these factors have changed to permit regeneration and survival of susceptible plant species and the change in vegetation in the presence of *P. cinnamomi*. Another question is whether regenerating susceptible species will be able to outcompete the dense flora comprising the understorey of infested sites, or whether the regeneration of susceptible plant species will stimulate the pathogenic activity of *P. cimmamomi* to reproduce epidemic and cyclic disease in suitable seasons. – However, it should be kept in mind that this is a rather specific example of the break down of a plant-pathogen population because the pathogen involved, *P. cinnamomi*, exhibits an extremely wide host range, which is rather uncommon with phytopathogenic fungi and oomycetes.

The nature of the resistance of wild plant populations that can be overcome so rapidly has not yet been analyzed. It might depend on a single component of the plant's basic resistance, and thus be vulnerable to the pathogen acquiring by mutation and/or recombination the ability to produce a pathogenicity factor required to overcome the "last" resistance component, or "residual basic resistance" of the plant. Residual basic resistance may be race specific, although it does not conform to a gene-for-gene interaction. Whether such resistance types can be enriched in wild populations has not yet been investigated. Analyzing the nature of such resistance determinants is worthwhile, because they might be more stable than race-specific resistance that relies on the gene-for-gene interaction.

The damage caused by phytopathogenic fungi to susceptible wild plant populations is generally less serious than that to populations of cultivated plants. This may be attributed to the pathogen's enormous potential for propagation on genetically uniform plants grown in single-crop farming and the high infection pressure resulting from it. By contrast, the lower genetic uniformity of wild populations results in variation in the susceptibility of individual host plants. Differences in residual basic resistance may also contribute to this variation.

Interactions between pathogens and host plants are in a state of permanent evolution with respect to both race-specific and other resistance mechanisms. Although the consequences depend on external factors such as climate and nutrition however, internal genetic factors in plants and pathogens, are of greater importance. Because both partners are subject to evolutionary change, new possibilities for interaction constantly emerge between them that contribute to either disease or resistance.

Reviews

Burdon J.J. (1993): The structure of pathogen populations in natural plant communities. Annu.Rev.Phytopathol. **31**: 305–323

Burdon J.J. (1997): The evolution of gene-for-gene interactions in natural pathosystems. In: Crute I.R., Holub E.B., Burdon J.J. (eds.): The Gene-for-Gene Relationship in Plant-Parasitic Interactions. CAB International, Oxon UK, New York NY. 245–262

Clarke D.D. (1997): The genetic structure of natural pathosystems. In: Crute I.R., Holub E.B., Burdon J.J. (eds.): The Gene-for-Gene Relationship in Plant-Parasitic Interactions. CAB International, Oxon UK, New York NY. 231–243

Frank S.A. (1992): Models of plant-pathogen coevolution. Trends Genet. **8**: 213–219

Frank S.A. (1993): Coevolutionary genetics of plants and pathogens. Evolutionary Ecology **7**: 45–75

Heath M.C. (1987): Evolution of plant resistance and susceptibility to fungal invadors. Can. J. Plant Pathol. **9**: 389–397

Heath M.C. (1991): Evolution of resistance to fungal parasitism in natural ecosystems. New Phytol. **119**: 331–343

Thompson J.N., Burdon J.J. (1992): Gene-for-gene coevolution between plants and parasites. Nature **360**: 121–125

Relevant papers

Burdon J.J., Wennström A., Elmqvist T., Kirby G.C. (1996): The role of race specific resistance in natural populations. Oikos **76**: 411–416

Weste G., Kennedy J. (1997): Regeneration of susceptible native species following a decline of *Phytophthora cinnamomi* over a period of 20 years on defined plots in the Gampians, Western Victoria. Austral.J.Bot. **45**: 167–190

9.1.8
Deviations from Strict Gene-for-Gene Relationships

Race-specific resistance in cultivated plants was the first system in which the genetic basis of host-pathogen interactions was experimentally analyzed. These investigations were conducted using different race-specific resistant cultivars, produced by plant breeding, and pathogen races isolated from nature with different specificities in specific virulence against the cultivars in use. Since Flor established the gene-for-gene hypothesis, only in a few other experimental systems have the genetics of race-specific resistance been thoroughly investigated and evidence obtained for a strict gene-for-gene interaction, i.e., that only one gene in the host plant and one in the pathogen are engaged in releasing expression of race-specific resistance. Absolute proof of this relationship requires rigorous verification by extensive crossing experiments supported by molecular genetic investigations in both the host plant and the corresponding homologous pathogen. (Compilations of the classic genetic investigations with references can be found in Crute 1985 and Christ et al. 1987). Applying such strict criteria, data obtained from only one of the partners is insufficient for establishing a gene-for-gene relationship, as is Person's indirect quadratic check method (see Sect. 9.1.1). Only in a few cases has clear evidence been obtained for exceptions from strict gene-for-gene interactions; these data show that in one, or in both, partners more than a single gene may be involved in releasing expression of race-specific resistance (see Christ et al. 1987.)

As in Flor's classic experiments with *Linum usitatissimum* and *Melampsora lini*, crop plant cultivars employed for investigating the genetic basis of race-specific resistance originated from long-term breeding programs for disease resistance. These began with dominantly expressed resistance markers that are easy to screen among the progeny of crosses. It is possible that choosing such favorable features resulted in the selection of a of resistance markers exhibiting gene-for-gene relationships. Other, less convenient, markers that also determine race-specific resistance may not act in a gene-for-gene manner.

It is not known whether all interactions in wild populations that release race-specific resistance are of the strict gene-for-gene type. However, Flor's experiments with flax and flax rust were extended to 30 different loci for race-specific resistance. Each of these determinants was shown to depend on a single gene or allele (Pryor and Ellis 1993). The race specific resistance genes of flax are distributed among five recombinational groups (K,L,M,N,P) some of which harbor multiple alleles. In one of them there are 13 distinct alleles that have not been separated by crossing because of their very tight linkage and the small number of progeny obtained in the crosses. However, in the flax rust pathogen, *Melampsora lini*, 24 different mutations were identified as being specifically virulent (Lawrence et al. 1981a,b). These 24 alleles "mark" 24 different avirulence genes (see also Sect. 9.1.1). Nineteen of them, which belong to four linkage groups, were analyzed further by crosses. For 14 of these alleles avirulence proved to be a single dominant trait, since after intercrossing and selfing the F_2 progenies segregated in a ratio of 3 : 1 avirulent to virulent on the appropriate monogenic resistant host cultivars. (*M.lini* is a dikaryotic fungus, thus segregations from crosses appear similar to those of a diploid organism.) However, in the remaining five specific virulence types, a dominant suppressor gene interfered with expression to prevent avirulence, thus effecting dominant expression of virulence. Hence, specific virulence appeared as a dominant trait and not a recessive one as expected from a strict gene-for-gene interaction. This result emerged from F_2 progenies which segregated 3 : 13 avirulent to virulent on the appropriate resistant hosts. Further experiments revealed that the dominant suppresor gene, called *I*, which was either a single gene with multiple alleles, or a cluster of five closely linked genes not separable by crosses, was able to suppress the five avirulence genes (Jones 1988; Christ et al. 1987; Lawrence et al. 1981a,b). A similar dominant suppression of avirulence resulting in the dominant expression of virulence was identified in *Puccinia recondita* f.sp. *tritici* (Christ et al. 1987; Samborski and Dyck 1968) and in *Puccinia graminis* f.sp. *tritici* (Christ et al. 1987; Kao and Knott 1969) on wheat cultivars.

Recessive suppressor genes for avirulence have also been identified. However, in a dikaryon both recessive alleles *i/i* must be present for expression of virulence, as was demonstrated in *Puccinia graminis* f.sp. *avenae* (Green and McKenzie 1967), *P. recondita* f.sp. *tritici* (Haggag et al. 1973), and *Ustilago hordei* (Ebba and Person 1975).

The action of suppressor genes involved in blocking race-specific resistance may be explained by proposing a gene-for-*gene-pathway* relationship instead of a gene-for-gene relationship. This means that the avirulence product is synthesized via a short metabolic pathway with a few intermediate products (*Fig. 12*). In this case, the avirulence factor would not represent a primary gene product in the strict sense since it would be determined by several genes rather than only one (see also Sect. 9.1.6). *Figure 12* shows three genes in a pathogen, Avr_1, Avr_2, and Avr_3, which participate in the synthesis of avirulence factor A_3. Notice that in a dikaryotic mycelium all genes are present in duplicate since each parent contributes one haploid nucleus to a cell. As a consequence, the phenotype expressed by a dikaryotic mycelium is the same as that of a diploid mycelium of the same genetic constitution. Since genes Avr_1 and Avr_2 are synthesizing precursors A_1 and A_2, which are not involved in recognition by $Rcpt_3$, avirulence factor A_3 appears as the primary gene product of gene Avr_3, as one may at first glance deduce from the observed phenotype. (In the host plant, two corresponding resistance genes R_1 and R_2 could emerge by mutation which recognize gene products A_1 and

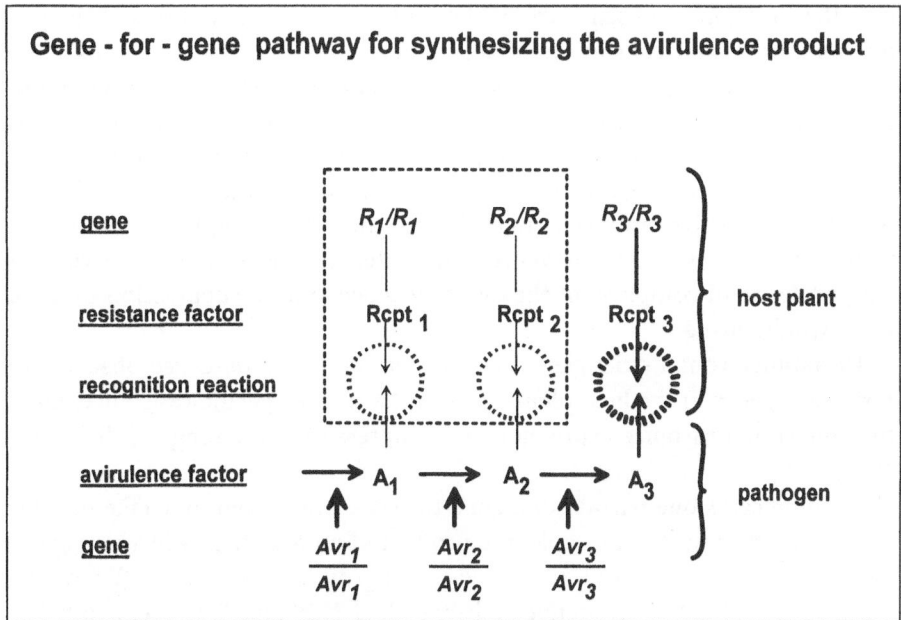

Figure 12. A gene-for-gene-pathway synthesizing avirulence factor A_3. A_3 is synthesized by the pathogen's dikaryotic mycelium within one pathway consisting of three genes or alleles, Avr_1/Avr_1, Avr_2/Avr_2, and Avr_3/Avr_3. Only avirulence product A_3 is recognized by the product $Rcpt_3$ of the diploid plant's resistance gene or its alleles R_3/R_3 thus releasing defense reactions. The other two gene products in the pathway, A_1 and A_2, do not correspond to the resistance factor $Rcpt_3$ and are not involved in recognition. For more details

A_2 encoded by avirulence genes Avr_1 and Avr_2. This would create two new gene pairs triggering release of defense reactions. This situation is shown inside a dotted square in *Fig. 12*. For the sake of simplicity, the present discussion will be confined only to recognition of A_3 by $Rcpt_3$.)

Dominant suppression of avirulence via the postulated gene-for-gene pathway could result from a mutational defect occurring within the pathway for synthesizing avirulence product A_3, for example, from Avr_2 to Avr_2^* in one of the parental nuclei of the dikaryon. The mutant allele Avr_2^* would correspond to the dominant suppressor gene *I*, and its allele *i* would correspond to the wild type allele Avr_2. Gene Avr_2^* would then synthesize product A_2^* acting as a suppressor and preventing recognition of resistance factor $Rcpt_3$ (*Fig. 13*, "either"). Alternatively, the A_2^* product could be converted by action of gene Avr_3 into product A_3^* which now also might serve as a dominant suppressor (*Figure 13*, "or"). The ensuing block of recognition by the suppressing products A_2^* or A_3^* would in both cases mask the resistance factor $Rcpt_3$ for example by competing successfully at $Rcpt_3$ for A_3. Blocking release of avirulence would be effective in the presence of a wild type Avr_2 allele since the A_2 product cannot complement the mutational defect of Avr_2^*.

A recessive suppression of avirulence may also be caused by a mutation within the pathway that directs synthesis of the avirulence product. However, in this case the functional consequence of the mutation causing suppression is quite different from that causing dominant suppression. For example, a mutation from Avr_2 to avr_2 might cause a complete loss of function because either no product is synthesized or the product is defective. The mutant allele avr_2 would correspond to the recessive suppressor allele *i*, and Avr_2 to *I* . Only if both recessive alleles, avr_2/avr_2 (or *i/i*), are present, either no A_2 product would be synthesized or only the defective product a_2. In both cases the substrate A_2 for synthesis of A_3 would be missing (*Fig. 14*). However, in the heterozygote avr_2/Avr_2, the wild type Avr_2 allele could complement the defective allele and no suppression of avirulence would ensue.

Deviations from a strict gene-for-gene determination were also observed for resistance genes in the host plant, e.g., recessive race-specific resistance, modifiers boosting resistance expression and suppressors preventing it (Christ et al. 1987).

In some cases one partner of a gene-for-gene interaction may take part in a second interaction. For example, the product of an *R* gene in wheat recognizes avirulence factors from one race of *Puccinia graminis* and one of *P. recondita* (McIntosh 1977). The respective resistance loci *Lr20* and *Sr15* seem to belong to the same resistance gene since they can not be separated in crosses and mutagenic treatment with EMS inactivates both loci simultaneously. However, a likely alternative interpretation would be that both pathogen species carry the same avirulence gene.- A comparable deviation from a strict gene-for-gene relationship was demonstrated for the avirulence factors of two different avirulence

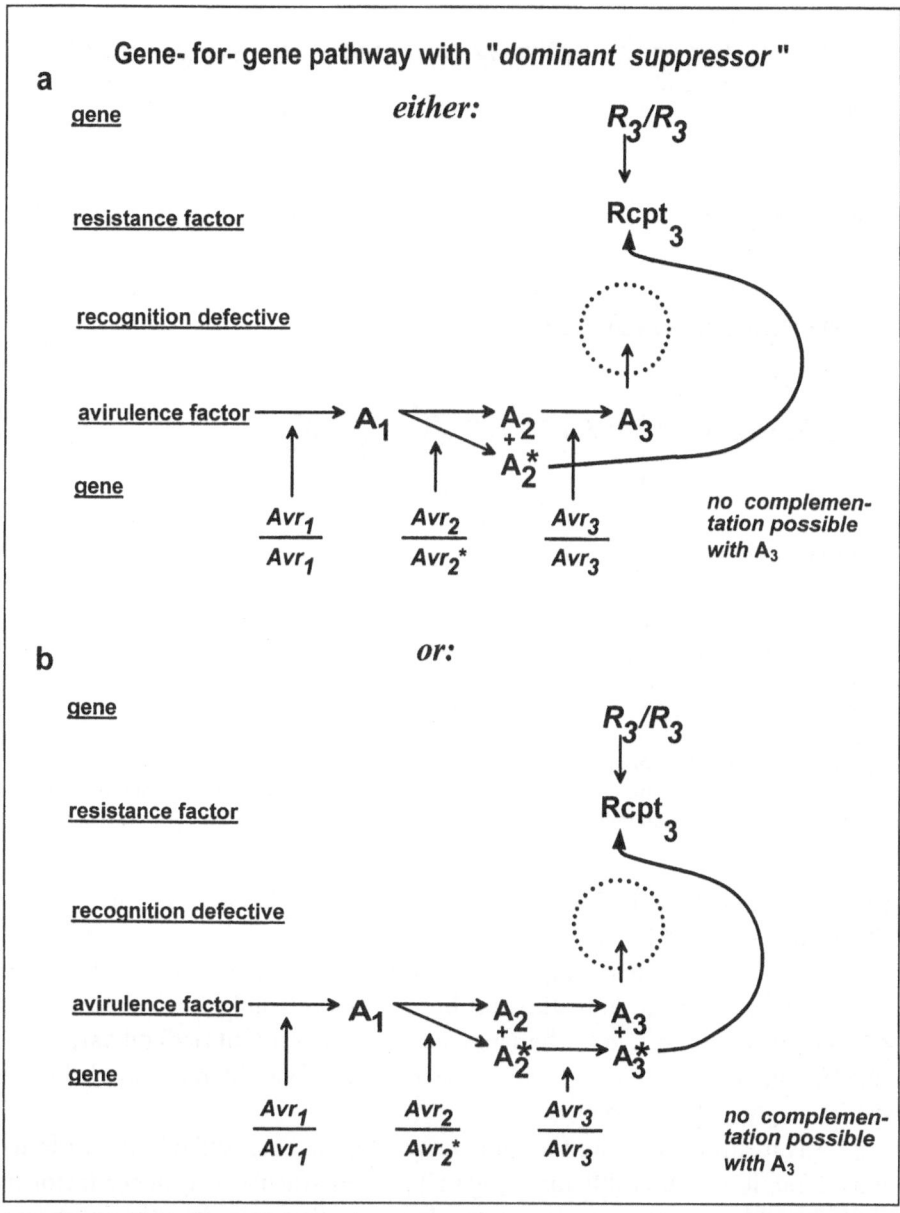

Figure 13a,b. The recognition between resistance and avirulence products is blocked by a dominantly expressed mutation within the pathway for synthesis of the avirulence product. Mutation of gene Avr_2 to the dominant suppressor gene Avr_2^* effects synthesis of suppressor product A_2^*, which is now synthesized in addition to A_2. Either the A_2^* product acts by itself as a suppressor (*13a*) or the A_2^* product is converted by the action of alleles Avr_3/Avr_3 to A_3^* (*13b*). Thus, either A_2^* or A_3^* could act as a suppressor by blocking the recognition reaction at Rcpt$_3$, the product of race-specific resistance gene R_3

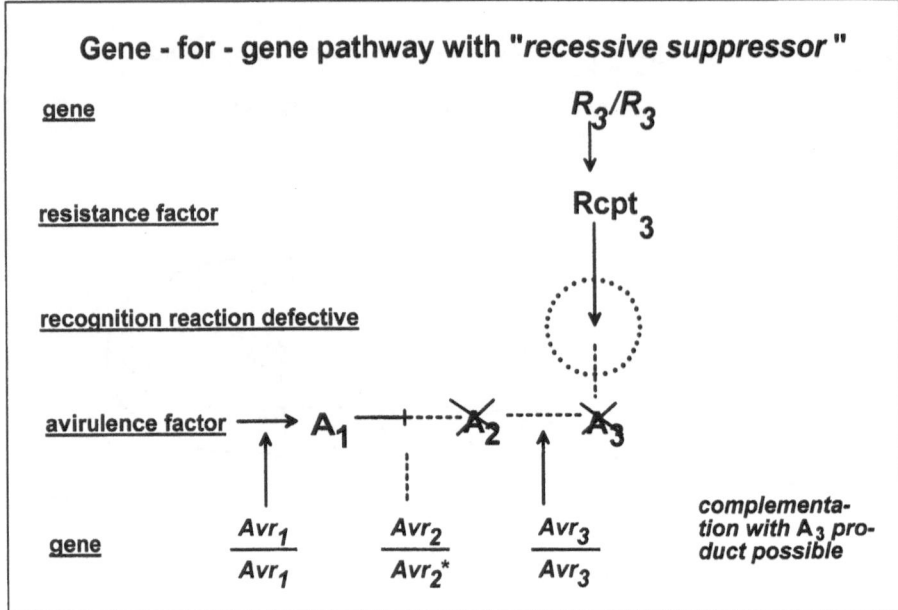

Figure 14. The recognition between resistance and avirulence products is blocked by a recessive mutation within the pathway for synthesis of the avirulence product. Mutations within alleles Avr_2/Avr_2 to avr_2/avr_2 prevent formation of a functional A_2 product. This interrupts the pathway to avirulence product A_3 and prevents a recognition reaction at $Rcpt_3$, the product of race-specific resistance gene R_3. Notice, that the recessive character of this suppressor mutation requires that both alleles are non-functional, otherwise a functional allele could complement a non-functional one

genes in *Ustilago hordei*, which are both recognized by the same *R* gene in barley (Ebba and Person 1975). In this case also two identical avirulence genes might be present in *U. hordei* originating from a chromosomal duplication-translocation. In *Magnaporthe grisea*, four different but very closely linked avirulence genes (see below) are recognized by the same race-specific resistant rice cultivar, called Katy (Ellingboe 1996). Katy was shown later to carry four different, closely linked resistance genes (see below).

As was mentioned earlier the same locus of an oat cultivar fulfills two functions. First, it is responsible for recognizing one particular avirulence factor of the biotrophic pathogen *P. coronata*, thus releasing race-specific resistance. Second, it determines sensitivity to the host-selective toxin victorin C produced by the necrotrophic pathogen *Cochliobolus victoriae* (*Helminthosporium victoriae*) (Rines and Luke 1985), which kills sensitive host plant cells (see also Chap. 8, Sect. 8.2 and Chap. 9, Sect. 9.1.3.3). Both functions, assigned to locus *Vb/Pc-2*, could not be separated by crossing. Furthermore, a detailed analysis of 155 lines with mutations that led to full resistance to victorin C also exhibited alterations in race-specific resistance against *P. coronata*. Of 41 mutants, 34 with only partial resistance to victorin C were also less resistant to *P. coronata* (Ellingboe 1976). It was thus hypothesized that locus *Vb/Pc-2* might be transcribed into two over-lapping transcripts, one being translated into the receptor protein for recognizing the avirulence product and the other into the receptor for recognizing victorin C (Ellingboe 1976).

Some gene-for-gene relationships may in reality turn out to be quite compli-cated as the next example demonstrates. In an analysis of rice cultivars carrying race specific resistance to blast, caused by the ascomycete *Magnaporthe grisea*, crosses confirmed that a single gene determined resistance in the cultivar Katy (Ellingboe 1992, 1996; Lau et al. 1993; Lau and Ellingboe 1993). Furthermore, crosses among *M. grisea* strains suggested that avirulence and specific virulence on Katy also depended on a single gene. These results appeared particularly trustworthy, since in this system both compatible and incompatible infection responses can be screened on the same host plant by the formation of large and of small infection lesions. Indeed, the following criteria argued in favor of a direct gene-for-gene relationship: (1) Crosses between two pathogen races virulent on cultivar Katy produced only virulent progeny, suggesting that only one gene is responsible for determination of specific virulence and avirulence. Correspondingly, (2), in crosses between strains avirulent on Katy, only avirulent progeny were produced. (3) Crosses between avirulent and virulent strains yielded segregation ratios of 1 : 1 for avirulence to virulence.

However, there was one conspicuous point: among the progeny of the crosses described in (3), the expression of avirulence and virulence appeared to be vari-able, as shown by the morphology of the infection responses. To analyze the ge-netic basis of this variability, some variable progeny pathogens were intercrossed and their progeny was tested again on Katy. In the progeny of some of these crosses phenotypes were found that contradicted a strict gene-for-gene relation-ship. For example, avirulent pathogens arose sporadically in the progeny of crosses between virulent strains, suggesting that the avirulence gene was still pre-sent, at least in one of the parental virulent pathogens, but was prevented from being expressed. This situation is equivalent to the suppression of avirulence, as discussed above for dikaryotic pathogens (*Figs. 13, 14*). However, *M. grisea* is haploid and it is not possible to discriminate between dominant and recessive suppression without further experiments. Furthermore, the avirulence and sup-pressor genes turned out to be very closely linked, and difficult to separate by recombination in crossing experiments. Segregation analysis of crosses among pathogens with varying phenotypes for avirulence and virulence led finally to the demonstration of 15 different avirulence determinants, each of which was closely linked with a specifically acting suppressor gene. Accordingly, avirulence is ex-pressed only when suppression is inactive (allele *s*), and virulence when suppres-sion is active (allele *S*).

The question arose of whether there is more than one suppressor gene that is specific for a particular avirulence gene. A blast culture of the genetic constitu-tion *P11S11 P12s12*, avirulent on Katy because of the gene pair *P12s12*, was treated with the mutagen nitrosoguanidine. Following mutagenesis virulent variants appeared carrying the suppressor mutations for avirulence, M_1^* and M_2^*. However, neither were linked to $P12s12$. If the genetic determination of avirulence/virulence depends, as in this case, on several genes, it becomes problematic to identify the gene actually encoding the avirulence product in order to clone and analyze its function. Only a very detailed genetic analysis should precede such a project. The question of whether suppressor genes S, M_1^* and M_2^* have other functions, remains unanswered.

Establishing how many race-specific resistance genes are carried by the cultivar Katy turned out to be more complicated than originally supposed. The early experiments suggested that only one resistance gene determined race-specific resistance in Katy. However, the unusual stability of this resistance and the presence of 15 different avirulence alleles in the pathogen (Ellingboe 1996) raised the suspicion that several resistance genes might be present. To answer this question an avirulent pathogen, Tm4, was isolated from an area where Katy is cultivated. (The avirulent characteristics of Tm4 on Katy correspond to a "slow rusting" phenotype (see Sect. 9.1), i.e., very slight colonization and reproduction of Tm4 spores on the resistant host plant. This resistance phenotype permitted the isolation of avirulent strains from Katy.) Crossing race Tm4 with a race virulent on Katy yielded segregation of 32 avirulent : 2 virulent progenies on Katy suggesting 4 genes determining avirulence on Katy. Pathogen recombinants containing only one avirulence gene were obtained by backcrossing avirulent progenies to a virulent race for several generations until the recombinant progenies segregated 1 avirulent : 1 virulent phenotypes. Intercrosses between these "single gene" strains showed that at least 4 different genes, each of which is capable of giving avirulence on Katy, were present in Tm4. Correspondingly, Katy should carry four different resistance genes. Indeed, crosses of Katy with other cultivars, and tests of the progeny plants obtained with pathogen strains of different race specificity, revealed that the four resistance genes proposed in Katy map closely linked within a cluster of at least six resistance genes. This clustering of resistance genes was obviously the reason why in the early experiments only one gene was identified as determining race-specific resistance. This detailed analysis strongly suggests that each of the four identified avirulence genes in Tm4 is involved in releasing race-specific resistance through a strict gene-for-gene interaction with the resistance determinants of Katy which map within one cluster. However, the presence of suppressor genes complicates the identification of the avirulence genes responsible for the recognition specificity.

The examples of proposed deviations from strict gene-for-gene relationships for the release of race-specific resistance discussed in this section confirm the general validity of the gene-for-gene principle. Interactions that at first glance do not strictly follow the gene-for-gene type may arise as a result of different, secondary, superimposed processes such as suppression. However, this does not exclude the possibility that some cases of apparent gene-for-gene relationships are actually based on much more complex processes. Since detailed genetic analysis of race-specific resistance has been carried out on only a few systems, the question remains whether strict gene-for-gene interactions really represent the dominant mechanism of race-specific resistance or whether they are an artefact of plant breeding.

Reviews

Christ B.J., Person C.O., Pope D.D. (1987): The genetic determination of variation in pathogenicity. In: Wolfe M.S., Caten C.E. (eds.): Populations of Plant Pathogens: Their Dynamics and Genetics. Blackwell Scientific Publications, Oxford. 7 – 19

Crute I.R. (1985): The genetic bases of relationships between microbial parasites and their hosts. In: Fraser R.S.S. (ed.): Mechanisms of Resistance to Plant Diseases. Martinus Nijhoff/Dr.W.Junk, Dordrecht, Boston, Lancaster. 81 – 142

Durbin R.D. (1983): The biochemistry of fungal and bacterial toxins and their modes of action. In: Callow J.A. (ed.): Biochemical Plant Pathology. John Wiley & Sons, Chichester. 137 – 162

Ellingboe A.H. (1976): Genetics of host-parasite interactions. In: Heitefuss R., Williams P.H. (eds.): Encyclopedia of Plant Physiology, (NS), Physiological Plant Pathology. Springer Verlag, Heidelberg. 761 – 778

Ellingboe A.H. (1996): Gene interaction in hosts and pathogens. In: Mills D., Kunoh H., Keen N.T., Mayama S. (eds.): Molecular Aspects of Pathogenicity and Resistance: Requirement for Signal Transduction. The American Phytopathological Society, St. Paul, Minnesota. 33 – 46

Gabriel D.W., Rolfe B.G. (1990): Working models of specific recognition in plant-microbe interactions. Annu.Rev.Phytopathol. 28: 365 – 391

Pryor T., Ellis J. (1993): The genetic complexity of fungal resistance genes in plants. Advances in Plant Pathology 10: 281 – 305

Relevant papers

Ebba T., Person C. (1975): Genetic control of virulence in Ustilago hordei. IV. duplicate genes for virulence and genetic and environmental modification of a gene-for-gene relationship. Can.J.Genet.Cytol. 17: 631 – 636

Ellingboe A.H. (1992): Segregation of avirulence/virulence on three rice cultivars in 16 crosses of Magnaporthe grisea. Phytopathology 82: 597 – 601

Green G.J., McKenzie R.I.H. (1967): Mendelian and extrachromosomal inheritance of virulence in Puccinia graminis f.sp. avenae. Can.J.Genet.Cytol. 9: 785 – 793

Haggag M.E.A., Samborski D.J., Dyck P.L. (1973): Genetics of pathogenicity in three races of leaf rust on four wheat varieties. Can.J.Genet.Cytol. 15: 73 – 82

Jones D.A. (1988): Genetic properties of inhibitor genes in flax rust that alter avirulence to virulence on flax. Phytopathology 78: 342 – 344

Kao K.N., Knott D.R. (1969): The inheritance of pathogenicity in races 111 and 29 of wheat stem rust. Can.J.Genet.Cytol. 11: 266 – 274

Lau G.W., Chao C.T., Ellingboe A.H. (1993): Interaction of genes controlling avirulence/virulence of Magnaporthe grisea on rice cultivar Katy. Phytopathology 83: 375 – 382

Lau G.W., Ellingboe A.H. (1993): Genetic analysis of mutations to increased virulence to Magnaporthe grisea. Phytopathology 83: 1093 – 1096

Lawrence G.J., Mayo G.M.E., Shepherd K.W. (1981a): Interactions between genes controlling pathogenicity in the flax rust fungus. Phytopathology 71: 12 – 19

Lawrence G.J., Shepherd K.W., Mayo G.M.E. (1981b): Fine structure of genes controlling pathogenicity in flax rust Melampsora lini. Heredity 46: 297 – 313

McIntosh R.A. (1977): Nature of induced mutations affecting disease reaction in wheat. In: Mike A. (ed.): Induced Mutations against Diseases. IAEA-SM-214/46, Vienna. 551 – 565

Rines H., Luke H.H. (1985): Selection and regeneration of toxin-insensitive plants from tissue cultures of oats *Avena sativa* susceptible to *Helminthosporium victoriae*. Theor. Appl.Genet. **71**: 16–21

Samborski D.J., Dyck P.L. (1968): Inheritance of virulence in wheat leaf rust on the standard differential wheat varieties. Can.J.Genet.Cytol. **10**: 24–32

9.2
Race-Non-Specific, or Horizontal, Resistance

There are two reasons why we discriminate between the two types of host resistance, race non-specific and race-specific resistance. First, the defense mechanisms are directed against different spectra of pathogens, either all the members of a species, or a *forma specialis* (race non-specific), or only to particular races or mutants (race-specific). Second, they differ in their genetic stability. Vanderplank called race-specific resistance "vertical resistance", whereas for the race-non-specific type he coined the term "horizontal resistance". He derived these terms from the responses he observed when infecting resistant host plants with various races of a pathogen species: He found either somewhat varying and more or less intermediate resistance effective against all the races ("horizontally" distributed resistance), or high resistance against one race and full susceptibility to others, ("vertically" distributed resistance) (see also Chap. 5).

Race-specific resistance is very effective but genetically unstable, and restricted to those pathogen races that carry a "corresponding" avirulence gene (for details see Sects. 9 to 9.1.2). By contrast, horizontally resistant plants defend themselves against every race of a particular pathogen. Horizontal resistance is genetically non-conditional, and very stable, but its efficacy is not as high as that of vertical resistance. Hence, it is also called partial resistance. However, the degree of horizontal resistance is influenced by environemntal factors such as humidity, temperature, and nutritional status. This type of host resistance will be treated below in more detail.

The genetic stability of race-non-specific or horizontal resistance results from two features: First, race-non-specific resistance cannot be overcome by a single mutational step in the pathogen, as occurs in the mutation to specific virulence observed in race-specific resistance (see Sect. 9.1.5). Rather, several mutations would be required in the pathogen to overcome host defense. Second, race-non-specific resistance is polygenically determined in the plant, each gene contributing only a small amount to the finally expressed phenotype. Because of its high genetic stability, breeding for horizontal resistance has become a primary goal in establishing biologically designed and environmentally adapted plant protection in agricultural production. This will no doubt be aided by techniques such as marker assisted selection which allow the selective accumulation of chromosome regions that contribute to resistance. However, identification and breeding for horizontal resistance is more complicated and expensive than breeding for race-specific resistance. Because of the experimental difficulties, a thorough analysis of the biology and genetics of race-non-specific resistance did not start much before 1960 and was pioneered by Vanderplank, Person, Parlevliet, Robinson and Zadoks.– Both designations, vertical resistance for race-specific and horizontal resistance for race-non-specific resistance, are used in the scientific literature. However, when referring to their genetic characteristics, the designations race-specific and race-non-specific resistance are to be preferred.

One way to distinguish race-specific from race-non-specific resistance is to compare the degree of plant resistance (or host injury) by inoculating cultivars of unknown resistance type with a set of different pathogen races. If all the different cultivar/pathogen pairs formed result in intermediate and varying degrees of resistance, i.e. showing neither complete susceptibility nor full resistance and if the levels of resistance (or degrees of injury) among the different cultivar/pathogen pairs are always ranked in the same order, this is called "constant ranking" (see Table 11, part A). Constant ranking is a characteristic feature of horizontal resistance which indicates the level of resistance expression depending mainly on the afflicted cultivar whereas the infecting pathogen race exerts only limited influence.

By contrast, if after infection of resistant plants with the set of different pathogen races either heavy colonization or no infestation at all is to be found, in other words the host plants exhibit either full susceptibility or high resistance but no intermediate responses, then the infected cultivars carry race-specific or vertical resistance. These extreme modes of infection responses – full susceptibility or a high level of resistance – indicate that the partners, plant and pathogen, contribute equally to the release of resistance reactions. The elements cooperating in this response are, on the plant side, a particular resistance gene (R_a), and, on the pathogen side, a particular avirulence gene (Avr_a); together these two genes, form a so-called corresponding gene pair (see Sect. 9.1.1). This mode of host-pathogen interaction has been called a "differential interaction". This is illustrated in Table 11, part B, showing either maximal damage to the host plant (grade 6 equaling full susceptibility) or no damage (grade 0 equaling full resistance). However, in practice, performing such "ranking order tests" for discriminating horizontal and vertical resistance requires considerable experimental effort.

A tentative distinction between race-specific and race-non-specific resistance may be obtained by measuring the levels of resistance of the same cultivars grown at various geographical locations under different conditions, all cultivars inoculated with the same tester pathogen races. Cultivars grown at different locations and still exhibiting the same rank order of resistance against the same set of tester pathogen races are most probably race-non-specific or horizontally resistant.

Race-non-specific resistance can also be identified by employing statistical tests of variance. If the amounts of injury caused by different pathogen races

Table 11. Constant ranking of race-non-specific (horizontal) and differential ranking of race-specific (vertical) resistant cultivars

A		Cultivars exhibiting **horizontal** resistance show *"constant ranking"*		
		D	E	F
Pathogen races	α	3	4	5
	β	2	3	4
	δ	1	2	3

B		Cultivars exhibiting **vertical** resistance show *"differential interaction"*		
		A	B	C
Pathogen races	α	6	0	0
	β	0	6	0
	δ	0	0	6

Each table shows idealized grades of damage observed after infecting two different sets of cultivar-specific resistant plants, one with horizontal, the other vertical resistance. In both tables, the infecting pathogen races are α, β, and δ. 0, complete resistance; 6, full susceptibility. – **A** contains horizontal or race-non-specific resistant cultivars D, E, and F. These cultivars exhibit neither full susceptibility nor complete resistance to the three pathogen races α, β, and δ and they always show the same ranking order in grades of resistances exhibited. This behavior is called "quantitative resistance" or "constant ranking". – **B** contains vertical or race-specific resistant cultivars A, B, and C showing either full susceptibility or complete resistance but no intermediate responses between the extreme reactions. This behavior is called "qualitative resistance" or "differential interaction".

infecting one particular race-non-specific (horizontally) resistant cultivar are compared, no statistically significant differences will be found. However, by employing the same test of variance with a race-specific (vertically) resistant cultivar, highly significant differences in the amounts of damage caused will appear among the different pathogen races. This is because the cultivars exhibit, depending on the infecting pathogen race, either high resistance or full susceptibility, i.e., an all-or-none reaction. Discrimination between horizontal and vertical resistance is also feasible when carrying out experiments in the opposite situation, namely, comparing the amounts of injury caused by a particular pathogen race infecting different horizontally or vertically resistant cultivars.

The characteristic differences in appearance of horizontal and vertical resistance may be summarized as follows: (1) Race-non-specific resistance is directed against all races or pathotypes of one pathogen or *forma specialis*, and not, as race-specific resistance, only against particular races. (2) Race-non-specific resistance is polygenically determined and results in only partial resistance in which defense against the pathogen is less complete than observed for race-specific resistance. Race-non-specific resistance has also been called quantitative resistance (see Sect. 9) in contrast to race-specific resistance which is qualitative and determined by one or a few major plant genes (see Sect. 9). (3) Because of its polygenic determination race-non-specific resistance, in contrast to race-specific resistance, is a very stable genetic trait. (4) Finally, expression of race-non-specific resistance does not require a corresponding gene in the attacking pathogen and therefore is not a conditional genetic trait, as is race-specific resistance. This non-conditional phenotype can be demonstrated either by a test of variance with respect to the resistance level of different horizontally resistant cultivars or by the constant ranking of different pathogen races when infecting different horizontally resistant cultivars. Table 12 summarizes the major differences between the two forms of resistance.

The terms minor and major genes, applied to genetic determinants of race-non-specific and race-specific resistance, do not assume different kinds of genes with different modes of expression. Rather, these terms were coined in classical genetics to designate genetic determinants engaged in either polygenic (several minor genes) or in mono- or oligogenic (one single or a few major genes) traits.

A gene determining race-specific resistance like R_a is expressed conditionally in the plant and mutations either in this gene, or in the corresponding pathogen gene Avr_a, may cause loss of resistance. Because of the high mutation rate of avirulence genes to specific virulence ($Avr \rightarrow avr$; see Chap. 9, Sects. 9.1.4, 9.1.5, and 9.1.6) combined with the high propagation rate and short generation time of pathogens, race-specific resistance appears to be a genetically unstable trait compared to race-non-specific resistance. Because farmers usually grow genetically uniform, race-specific resistant plants of single-crops, any spontaneous specific virulent mutant in the attacking pathogen population will be selected and will multiply very efficiently. As a result, the previously race-specific resistant cultivar may, after several seasons, become susceptible to the attacking pathogen population, which consists now mostly of self-selected specific virulent mutants. In contrast, a race-non-specific resistant cultivar would exhibit a ge-

Table 12. The two types of cultivar-specific resistance: race-non-specific (horizontal) and race-specific (vertical) resistance of host plants against colonization by phytopathogenic fungi: Comparison of some properties of both kinds of host resistance

	Horizontal resistance	Vertical resistance
Genetic determination	Polygenic	Monogenic or oligogenic
Gene action	Additive ("quantitative")	All or none ("qualitative")
Specificty	Species-specific	Race-specific
Correspondence between genes in host and pathogen	Absent	Present
Durability of resistance	Stable	Unstable
Designation of participating genes	Minor genes	Major genes

netically very stable pathogen defense because this type of host resistance is non-conditional and cannot be overcome by a single mutation in the pathogen to specific virulence. Since it is polygenically determined overcoming, or bypassing, only one of the many plant resistance determinants by a compensating mutation in the pathogen would result only in a negligible reduction of expressed resistance.

The polygenic determination of race-non-specific resistance can be demonstrated by crossing experiments: After crosses between susceptible and horizontally resistant cultivars, only a few progeny plants exhibit in the F_2 generation either the full susceptibility or the original race-non-specific resistance of the parent plants. The other segregants will show a normal distribution around the mid-parental mean confirming the presence of many independently segregating resistance genes, each gene contributing a small amount to the overall resistance and segregating in the F_2 progeny. In contrast, crosses between a monogenically determined, race-specific resistant cultivar and a susceptible cultivar would yield in the F_2 generation a discontinuous distribution of plant phenotypes, namely, exclusively parental phenotypes exhibiting either full resistance or full susceptibility (see also *Fig. 3*), masking the effects of any minor genes that may also be segregating. The mechanisms of race-non-specific resistance and how its expression is regulated is largely unknown because the contribution of each of the participating minor resistance genes to the overall horizontal resistance is very small and difficult to identify and quantify. For the same reason, there is also no experimental evidence as to whether minor gene action effecting pathogen defense relies on constitutive or induced expression, either by a gene-for-gene interaction or by any other mechanism.

Some attempts to answer these questions from theoretical model calculations show, for example, that in the presence of five minor genes for resistance the strength of race-non-specific resistance would be independent of their mode of action, either gene-for-gene or not. Race-non-specific resistance could, in principle, even rely exclusively on many single gene-for-gene interactions, each contributing a small amount to the overall resistance. In this case the additive gene action in horizontal resistance would be even more stable genetically than if each single contribution to resistance were determined non-conditionally, i.e., by resistance determined only by genes in the plant.

From such calculations race-non-specific and race-specific resistance, might both be integrated into a single hypothetical concept based solely on gene-for-gene interactions. In natural ecosystems, the equilibria between host plants and their pathogens that ensure the survival of both partners, would then depend on the product of the efficacy of each resistance gene multiplied by their frequency in the plant population. The efficacy of a single major resistance gene would be high but that of a minor gene would be low, and the frequency of major resistance genes would be low, with minor genes high. However, under the unnatural conditions of genetically homogeneous single-crop farming, the significance of major gene race-specific resistance has been increased enormously because of the strong selection it imposes for specific virulent pathogen mutants (*avr*). Thereby the distorted equilibrium between host plants and their pathogens favoured by the farmer is broken down.

The phenotypes of race-non-specific resistance are somewhat variable. For example, there are differences in the periods of reduced susceptibility of the plant to pathogen infection, that contribute to so-called disease escape. There is also variation in the reduction of pathogen penetration rate, which results in diminished numbers of productive lesions; the decrease in pathogen growth rates inside the plant with subsequent formation of smaller lesions; differential development of dwarfed fruiting bodies by the pathogen; reduced sporulation rate; extension of the pathogen's latent period (the time span from infection to spore production); and other properties that disadvantage the pathogen when infecting the plant. The slow rusting phenotype may also be caused by such plant-determined impairment of pathogen development. Clearly, such phenotypes depend on the action of many different plant genes. When breeding for horizontally resistant cultivars, only a few genetic markers can easily be selected for, such as decreased infection or sporulation rates. Other plant phenotypes, such as a changed susceptible infection period, are much more difficult to screen or select for.

While most resistance phenotypes are expressed during the entire lifetime of the plant, there are some interesting exceptions. Race-non-specific resistance against some biotrophic pathogens may be expressed only within a particular, ontogenetically determined period of plant development. For example, the resistance of some cereal cultivars to rusts or powdery mildews is expressed solely in adult plants and not in seedlings. Adult plant resistance (APR), like other types of race-non-specific resistance, may be based on decreased infection frequencies, increased latent periods for the pathogen, or shortened sporulation periods, and is determined by the additive actions of several minor genes. APR expression may be modified in certain regions of the plant body in time as well as in level, according to the developmental state of the plant. However, APR can also be modified by environmental conditions, e.g., temperature, humidity, light intensity, ultraviolet radiation, and similar microclimatic parameters.

Race-non-specific resistance may be improved by introducing additional genetic determinants through plant breeding. These may be either genes for race-specific resistance or determinants that increase resistance by unknown mechanisms, which, in the absence of horizontal resistance, have no detectable phenotype. However, horizontal resistance is not the only means of obtaining genetically stable, cultivar-specific resistance. For example the *Mlo* locus of barley (see Chap. 9, Sect. 9.1.4) confers genetically stable host resistance against all races of *E. graminis hordei*, and is determined by mutation in only one chromosomal locus or gene.

Resistant cultivars genetically stable over many decades of cultivation, in different regions, and under favorable infection conditions, show so-called "durable resistance" (Johnson 1981, 1984, 1992). Durable resistance may be determined by a single gene or by a combination of different genes, such as race-specific or race-non-specific resistance, or genes that increase resistance by unknown mechanisms.

Cultivars with horizontal resistance, or any other resistance of complex genetic determination such as durable resistance, were mostly selected within extensive breeding programs. These originated either by crossing an existing cultivar with a resistant wild relative, or by crossing distantly related cultivars, to introduce sufficient genetic variability. After several successive generations of selfing, or backcrossing to the desired susceptible parent, the most susceptible and most resistant plants were identified and discarded: the susceptible plants because they exhibited the wrong phenotype and the highly resistant plants because they might harbor an unwanted major gene determining race-specific resistance concealing the presence of determinants for horizontal or partial resistance. In cereal breeding likely race-specific resistant mutants can be eliminated by discarding resistant seedlings and subsequently screening the susceptible survivors for adult plant resistance.

In nature, the prototypes of our cultural plants existed together with their pathogens in some kind of equilibrium, kept in balance by genes in the plant population expressing host resistance. These plants were domesticated by humans who selected genotypes to create cultivars to meet their demands. At first their selective criteria on the whole preserved the existing natural equilibrium between plant and pathogen. However, during the last century, humans began large-scale efforts to breed high yielding cultivars with very specific characteristics. As a result, many genes participating in the expression of host resistance were most probably lost because of human plant-breeding efforts. This loss in genetic diversity especially affected genes involved in horizontal resistance.

However, the high yielding cultivars were susceptible to diseases which jeopardized their profitability. As a result resistance to infection based on race-specific or vertical resistance was increasingly emphasized. The selection of cultivars with corresponding major genes obtained by screening germplasm collections, or from spontaneous or induced mutagenesis, was easy because of their strong and dominantly expressed phenotype. However, this form of resistance soon proved to be unstable as specific virulent mutants appeared that were selected by the new cultivars. To maintain efficient pathogen defense the steady introduction of new cultivars carrying newly selected race-specific resistance genes resulted in a human-guided coevolution of race-specific resistant cultivars and specific virulent mutants among the homologous pathogen populations (see also Chap. 9, Sect. 9.1.5). Because of this coevolution for the last 25 years, efforts have been shifted to breeding for the somewhat less effective but genetically stable race-non-specific resistance.

Reviews

Boyle C., Schönbeck F. (1997): Durable resistance. In: Hartleb H., Heitefuss R., Hoppe H.-H. (eds.): Resistance of Crop Plants against Fungi. Gustav Fischer, Jena, Stuttgart, Lübeck, Ulm. 254–271

Burdon J.J. (1987): Diseases and Plant Population Biology. Cambridge University Press, Cambridge, London, New York, New Rochelle, Melbourne, Sidney

Crute I.R. (1985): The genetic bases of relationships between microbial parasites and their hosts. In: Fraser R.S.S. (ed.): Mechanisms of Resistance to Plant Diseases. Martinus Nijhoff/Dr.W.Junk, Dordrecht, Boston, Lancaster. 81–142

Ellingboe A.H. (1981): Changing concepts in host-pathogen genetics. Annu.Rev.Phytopathol. 19: 125–143

Johnson R. (1984): A critical analysis of durable resistance. Annu.Rev.Phytopathol. 22: 309–330

Johnson R. (1992): Reflections of a plant pathologist on breeding for disease resistance, with emphasis on yellow rust and eyespot of wheat. Plant Pathology 41: 238–254

Parlevliet J.E., Zadoks J.C. (1977): The integrated concept of disease resistance: A new view including horizontal and vertical resistance in plants. Euphytica 26: 5–21

Robinson R.A. (1971): Vertical resistance. Review of Plant Pathology 50: 233–239

Robinson R.A. (1973): Horizontal resistance. Review of Plant Pathology 52: 483–501

Simmonds N.W. (1991): Genetics of horizontal resistance to diseases of crops. Biol.Rev. 66: 189–241

Vanderplank J.E. (1963): Plant Diseases: Epidemics and Control. Academic Press, New York, London

Vanderplank J.E. (1968): Disease Resistance in Plants. Academic Press, New York, London

Vanderplank J.E. (1984): Disease Resistance in Plants. Second Edition. Academic Press, Orlando, San Diego, San Francisco, New York, London, Toronto, Montreal, Sidney, Tokyo, Sao Paulo

Watson I.A. (1970): Changes in virulence and population shifts in plant pathogenesis. Annu.Rev.Phytopathol. 8: 209–230

Zadoks J.C. (1972): Modern concepts of disease resistance in cereals. In: Lupton F.G.H., Jenkins G., Johnson R. (eds.): The Way ahead of Plant Breeding. Proc.6th Congr. Eucarpia, Cambridge. 89–98

Zadoks J.C. (1972): Reflections on disease resistance in annual crops. In: Bingham R.T., Hoff R.J., McDonald G.I. (eds.): Biology of Rust Resistance in Forest Trees. U.S. Department of Agriculture, Forest Service, Washington, D.C. 43–63

Relevant papers

Johnson R. (1981): Durable resistance: Definition of, genetic control, and attainment in plant breeding. Phytopathology 71: 567–568

When susceptible host plants are challenged with homologous phytopathogenic fungi, selection for resistant host plants is not the only possible outcome. Infection tolerant plants may also be selected. Infection tolerance occurs when the host plant, although infected and colonized by a phytopathogenic fungus, is not noticeably disadvantaged in its growth and reproduction when compared with an uninfected plant.

The distinction between tolerance and resistance was a subject of some controversy , since tolerance towards a pathogen was regarded as a special form of resistance. However, if resistance is defined as the capacity of a plant to defend itself against attacking pathogens and prevent colonization, tolerance against a homologous pathogen is a different phenomenon. Tolerance implies that the plant presents no defense reactions against the invading pathogen. Rather, the plant continues to thrive despite being colonized by compensating for the impairments, which might otherwise limit or even prevent its growth and reproduction. According to this definition, the phenomenon of slow rusting, observed with some rust-infected, race-specific resistant cereal cultivars, is not an example of tolerance since it depends on expression of plant defense reactions that bring about only incomplete resistance (see Sect. 9.1).

There are two forms of infection tolerance, passive and active. In passive tolerance the parasite develops inside the plant without producing any disease symptoms. A good example is the colonization of oat cultivars by the phytopathogenic fungus *Cochliobolus victoriae* (*Helminthosporium victoriae*), when no disease symptoms and no yield losses are observed. (However, oat cultivars containing the chromosomal locus determining sensitivity to the host-selective toxin victorin C produced by *C. victoriae* would be killed; see Chap. 8, Sect. 8.2). In active tolerance the host plant compensates for the drawbacks caused by pathogen infection by altering its own metabolic activities, for example by producing new leaves or adventitious roots.

The exploitation of infection tolerance for crop protection seems an obvious strategy, since tolerant plants are either not, or only minimally, impaired when infected. Also we would expect that tolerance would be genetically more stable than race-specific resistance. The unhindered pathogen reproduction in tolerant cultivars should impose little or no selection for pathogen mutants which reproduce faster, thus causing more damage to their host plants, than the wild type pathogen. So far, the development of pathogen-tolerant cultivars has been accidental. Selection for active infection-tolerance is feasible only with some understanding of the processes underlying the disease symptoms, and that enable the plant to compensate for the infection symptoms. This information should lead to reliable selection criteria for plant breeders interested in developing tolerant cultivars.

Infection-tolerant cultivars should never be used close to race-specific cultivars, i.e., within the same cultivation area. Under such circumstances the spontaneous appearance of specific virulent pathogen mutants, during pathogen growth on an infection-tolerant cultivar, would negate the race-specific resistance of the adjacent cultivars. Thus infection-tolerant cultivars may be a source of pathogen mutants, including those with specific virulence on certain race-specific resistant cultivars.

Reviews:

691 Clarke D.D. (1986): Tolerance of parasites and disease in plants and its significance in host-parasite interactions. Advances in Plant Pathology 5: 161–197
628 Schafer J.F. (1971): Tolerance to plant disease. Annu.Rev.Phytopathol. 9: 235–252

Host plants can be protected against further pathogen attack if they have survived earlier infection by phytopathogenic viruses, bacteria or fungi. Protection is also observed after an attack by herbivorous arthropods, mechanical injury or following contact with certain chemicals. It appears that the first infecting pathogen, or some an injury, "immunizes" the plant against further infections by homologous pathogens, even though the plant may not carry gene(s) determining cultivar-specific resistance. Obviously, the first infecting pathogen – or an injury – "induced" expression of resistance reactions against subsequently infecting pathogens, regardless of whether they are phytopathogenic viruses, bacteria or fungi. This "readiness" of the plant to repel subsequent pathogen attacks spreads throughout the whole plant. This response is called *systemic acquired resistance* (SAR). Protection by SAR may last for 6 weeks or longer. SAR is distinguished from race-specific resistance in that it decays with time and is effective against a broad spectrum of different pathogens. The spectrum of pathogens may vary with the plant species afflicted (Schönbeck et al. 1980; Moffat 1992). Thus injury caused by herbivorous insects, or by mechanical wounding, also induces expression of SAR, which then protects against attack by other feeding insects, but may not protect against attacking viruses, bacteria or fungi. The reverse may also be true: Primary infection with phytopathogenic viruses, bacteria or fungi may not protect against secondary injury by herbivorous insects.

The strength and stability of induced resistance over several weeks may be influenced by such factors as climatic conditions and nutrition. The observation that SAR spreads in the plant mainly in an apical direction, and moves into grafted stems, strongly suggests that signaling agents that establish SAR are translocated through the plant.

Induced resistance has been demonstrated in many cultivated plants including cucumber, melon, bean, tobacco, tomato, potato, grape, coffee, barley, wheat, pear, apple, plum and carnation. Further analysis has shown that a variety of defense mechanisms may be employed in SAR. For example, expression of SAR in cucumber against infecting *Colletotrichum lagenarium* has been demonstrated to be a result of hindrance of fungal hyphae penetration by formation of a callose-like papilla at the inner side of the epidermal cell wall below the appressorium. Also increased lignification of epidermal cells hinders penetration. Furthermore, green beans expressing SAR show increased levels of phytoalexin synthesis when challenged with *C. lindemuthianum*, and induction of SAR in tobacco against infecting *Peronospora tabacina* consists in accumulation of β-1,3 glucanases and chitinases. Generally, SAR expressing plant cells exhibit accumulation of PR-proteins. The recent demonstration of SAR in *Arabidopsis* offers the opportunity for a detailed molecular biological analysis.

An essential requirement for SAR is that the first infecting pathogen causes a necrotic lesion. This lesion may originate either from an incompatible interaction of a race-specific resistant cultivar with an avirulent pathogen, causing a hypersensitive reaction (HR) leading to the formation of a necrotic lesion or from a compatible interaction with accompanying necrosis formation. Whereas the connection between PCD, resistance expression, and the role of salicylic acid in it is well established (see Chap.9, Sect.9.1.4), the connection between PCD, and SAR is unclear, although also salicylic acid plays an important role here. Defense genes induced by a primarily infecting pathogen that are expressed locally as well as systemically in the plant are called SAR genes. Other plant genes that also govern defense reactions are not systemically expressed.

The sequence of events leading finally to SAR begins locally, i.e. close to the HR in adjacent cells with depositions of callose, thickening of cell walls by incorporation of structural proteins or lignin, and induction of phytoalexin synthesis. In the more distant non-infected parts of the plant the first of SAR-type defense reactions are the synthesis of PR-proteins (see Chap. 5), and the enzymes β-1,3 glucanases (Shah and Klessig 1996), endohydrolases, chitinases, enzyme inhibitors like thaumatins, amylase- and proteinase inhibitors. Unlike the highly specific recognition mechanisms that release race-specific resistance and effect HR and PCD, SAR is effective against a broad spectrum of homologous pathogen species and races.

Analysis of the induced resistance to insect herbivores showed that it is caused by plant-synthesized proteinase inhibitors, which block the digestive function of these insects. Signals responsible for spreading the release of synthesis of proteinase inhibitors were first investigated when the expression of SAR was analyzed. The responsible substance, which was translocated apically, was called proteinase inhibitor inducing factor (PIIF; Roberts 1992, Ryan 1992). PIIF was shown to be transported via the phloem. Furthermore, intracellular signals and regulatory substances were also identified that were engaged in activation of a gene called *pin*, coding for the *p*roteinase *in*hibitor, and of other genes participating in expression of defense reactions. Among the activating signals or regulatory substances were found ethylene; jasmonic acid and its volatile jasmonic acid methylester (Farmer and Ryan 1990), the latter both freed from the plasma membrane by lipase and other enzymes; abscisic acid; and systemin, a polypeptide of 18 amino acids. All of these substances could be secondary messengers within a long signal transduction chain effecting activation of defense reactions, or they are involved in signal transduction in some unknown way. Systemin seems to be the substance most close to the start of the signal transduction chain, and is possibly translocated systemically like a hormone, whereas abscisic acid and jasmonic acid are more downstream signal substances. All three signal substances have been identified also after mechanical wounding. – In general, signal transduction chains are made up of a sequence of different substances, each of which releases the synthesis of the next signal substance downstream. The last substance of the chain releasing a function, such as the transcription of a gene engaged in pathogen defense.

In similar investigations of SAR released after infection by bacteria or fungi salicylic acid (SA) was identified as a putative signal substance. SA is active in releasing SAR and synthesis of PR-proteins when applied experimentally externally (see below). Evidently the plant's repertoire of different signal substances for releasing SAR, as well as its repertoire of different defenses, allow a selective response that depends on the nature of the injury, either mechanical wounding, attack by herbivorous insects, or infection by phytopathogenic fungi or bacteria. There may also be some overlap among the defense responses, for example, proteinase inhibitors acting against enzymes produced by insects as well as against proteinases produced by bacteria and fungi.

SAR may also be released by spraying or injecting plants with salicylic acid, as mentioned above, or by treatment with H_2O_2, methyl-2,6-dichlorisonicotinic acid (INA), or benzo(1,2,3)thiadiazol-7-carbothionic acid (BTH) (Metraux et al. 1991; Ward et al. 1991; Uknes et al. 1992). These chemicals do not function as antibiotics by impairing the attacking pathogen; rather, they appear to induce in the plant expression of the same resistance response spectrum and the same set of SAR genes as is induced by an infecting pathogen. This strongly suggests that INA and BTH are true activators of SAR, whereas H_2O_2 (see below) and salicylic acid are metabolites present in the plant which are directly involved in establishing SAR. For example, in *Arabidopsis thaliana* it was demonstrated that infection by tobacco mosaic virus (TMV) as well as treatment with salicylic acid or INA released synthesis of the same set of 13 different PR-proteins. It was also shown that INA is translocated within 1 day from the upper leaves of young cucumber plants down into the roots, releasing there the synthesis of chitinase, i.e., a defense reaction against bacterial and fungal infections. However, it still remains to be shown whether INA and BTH act because of their steric relationship to salicylic acid or if they are more indirectly involved by formation, or activation of, other signal substances.

Salicylic acid plays an important role within the signal transduction chain releasing defense reactions in race-specific resistance. In Chap. 9, Sect. 9.1.4 was noted that in *nahG* transgene *Arabidopsis* lowering the intracellular level of salicylic acid, by its conversion into catechol, prevented expression of resistance against infecting *Perenospora parasitica* or *Pseudomonas syringae* (Delaney et al. 1994). Salicylic acid is also necessary for the expression of SAR. For example, after tobacco mosaic virus (TMV) infection tobacco (*Nicotiana tabacumum* cv. *xanthi*-nc, TMV susceptible) accumulates salicylic acid and expresses SAR in non-infected leaves. Both effects are blocked in corresponding *nahG* transgeneic plants which transform intracellular salicylic acid into catechol by salicylate hydroxylase thereby lowering the level of intracellularly accumulated salicylic acid. After challenge with TMV the size of the lesions that develop is inversely proportional to the amount of salicylic acid that has accumulated, since SAR expression is directly proportional to accumulated salicylic acid (Gaffney et al. 1993).

Interestingly, wild type tobacco scions express SAR even when grafted onto TMV-infected *nahG* stock plants free of salicylic acid (Vernooij et al. 1994). The authors concluded that the *nahG* stock plant produces, after TMV infection, some signal substance other than salicylic acid, which is translocated into the wild type scion, and releases the expression of SAR which is assisted by salicylic acid synthesis. Using salicylic acid produced in inoculated and healthy leaves and labeled there non-invasively with $^{18}O_2$, Shulaev et al. (1995) showed that after TMV infection (one basal leaf of *N. tabacum* cv. *xanthi*-nc, carrying the TMV resistance by gene *N*) the salicylic acid that accumulated in the upper uninfected leaves was 60 – 70 % ^{18}O labeled. It was concluded that salicylic acid is indeed the signal releasing SAR expression being exported from the infected leaf to healthy ones. In the experiments of Vernooij et al. (1994) possibly not all the salicylic acid synthesized in the *nahG* transgeneic rootstock after TMV infection was transformed into catechol and the low level of salicylic acid left was sufficient for systemic translocation via phloem to the upper leaves to release there the expression of SAR.

The signal transduction chain continues downstream of salicylic acid, before finally releasing expression of SAR genes. This is valid, within the framework of race specific resistance, for the expression of local defense reactions in cells immediately adjacent to necrotic race-specific HR as well as for the release of systemic SAR. This dual function of signal transduction is schematically shown in *Fig. 15*.

Salicylic acid is synthesized via phenyl propanoid metabolism from cinnamonic acid and benzoic acid. It accumulates intracellularly at a specific receptor or binding protein which has been identified as a catalase. Normally, catalase protects plant cells against the oxidative stress exerted by active oxygen compounds. However, this activity of catalase is blocked by the binding of salicylic acid (Chen et al. 1993). Thus, by altering the amount of salicylic acid inside the cell, the level of active oxygen compounds may be regulated: If only small amounts of salicylic acid are present in the cell, and there is little bound to catalase, the latter's activity will remain high, keeping the level of active oxygen low. When the level of salicylic acid is high and catalase activity low, the level of active oxygen will remain high thereby releasing synthesis of PR-proteins. Synthesis of PR-proteins can also be induced artificially by injecting H_2O_2 into plant leaves. The oxidative burst following pathogen infection can now take its optimal effect, as the accompanying increase in salicylic acid synthesis lowers catalase activity, permitting PR-protein expression triggered by the active oxygen compounds.

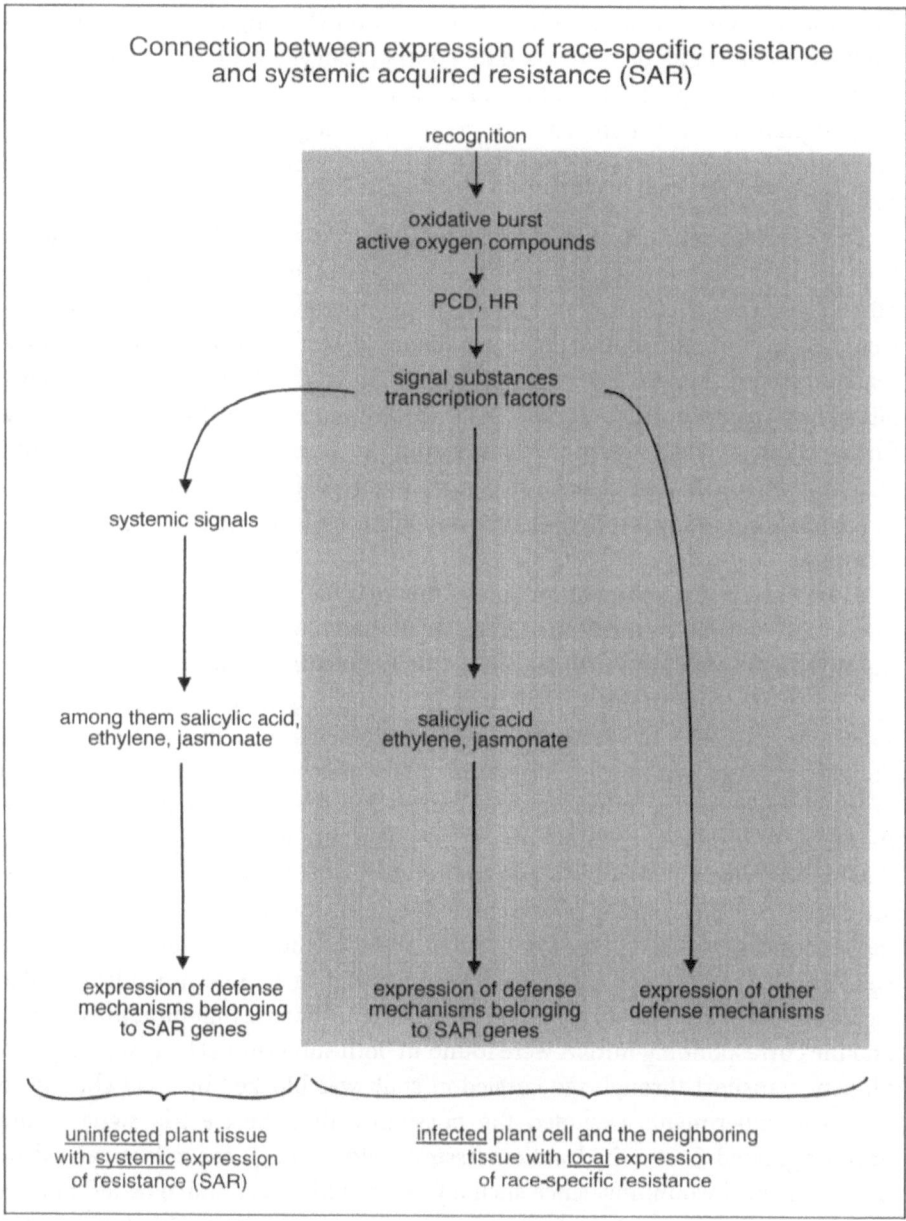

Figure 15. Connection between expression of race-specific resistance and systemic acquired resistance (SAR). SAR genes express defense-related proteins (enzymes, enzyme inhibitors) and pathogenesis-related(PR)-proteins. „Other defense mechanisms" include: callose deposition, cell wall thickening (lignin formation, structural proteins), and phytoalexin synthesis

Salicylic acid, and its precursor benzoic acid, may be removed intracellularly from this regulatory equilibrium by glucosylation (Hennig et al. 1993). The regulatory connections between oxidative burst, salicylic acid and expression of PR-proteins while plausible within single cells does not readily explain SAR release in plant tissue remote from the point of pathogen infection where signals other than salicylic acid are involved in releasing expression of SAR genes (see also *Fig. 15*).

The biochemical basis of the events in cell necrosis, are still largely unknown (*Fig. 15*). The only exceptions are the reactions associated with the oxidative burst (see Sect. 9.1.4). However, mutant phenotypes of *Arabidopsis thaliana* were recently identified that have regulatory defects in some of the putative reaction steps leading to cell necrosis. Among them are the *lsd* and *acd* mutants already discussed in Chap. 9, Sect. 9.1.4 (3). Mutants of this kind will facilitate further analysis of the reaction chain leading to necrotic race-specific HR and will help in identifying signal substances involved in SAR and, in particular, in translocation to uninfected plant tissue far away from the site of primary infection.

A very interesting observation about the possible nature of signal translocation in SAR should be mentioned here. As discussed above, mechanical wounding is thought to induce a signal, PIIF, which spreads systemically and releases the expression of proteinase inhibitors in uninfected plant tissue. In principle such a signal could be electrical in nature. Evidence that this indeed may be the case emerged from the observation that salicylic acid or fusicoccin, both of which impede ion transport, could affect the speed of signal transduction. Recently, Wildon and collaborators (1992), working with young tomato plants that had developed cotyledons, and up to two foliage leaves, demonstrated that, after mechanical wounding of one cotyledon, a short electric potential alteration moves as a signal through the symplast via the cotyledon stalk down to the main stem and from there upwards into the second and first subterminal leaves. The signal spread at a speed of 1 – 4 mm/s. After 48 h proteinase *in*hibitor, *pin* activity, and the corresponding mRNA were found in both subterminal leaves, even when phloem transport through the cotyledon stalk was blocked by local chilling to 3°C. The experiments excluded the possibility that the electric signal could have originated from a hydraulic pressure wave through the xylem caused by the mechanical wounding, since such a wave would spread much faster, at least 1000 mm/s. The experimental results point to a causal relationship between the observed systemic electric signal and the spread of proteinase inhibitor activity within the plant body. However, this mode of transmission would function only in very young plants. A similar phenomenon has been found in conduction systems of animals. In epithelial tissue of siphonophores and hydromedusae, which are free of muscle and nervous cells, and in the tissue of at least four other phyla of animals, electrical conduction has been demonstrated involving cell-to-cell contacts through gap junctions which takes place at about the same speed

(2–500 mm/s) as observed in the young tomato plants. The plasmodesmata that connect adjacent plant cells could function in this way may well facilitate the passage of electrical signals.

Induced systemic resistance has a very interesting practical aspect: It allows the plant to defend itself against different homologous pathogen species by the same mechanisms that are involved in releasing race-specific resistance. In this way mechanisms from the plant's repertoire of inducible defense reactions can be employed without requiring that the cultivar possess the corresponding race-specific resistance genes for pathogen defense. This implies, with admittedly some minor exceptions, that the expressed defense is directed non-specifically against all homologous pathogens, not only against a particular *forma specialis* or even only against certain pathogen races, as is the case in cultivar-specific resistance. However, pathogen defense by SAR has one drawback: Its efficiency is lower than pathogen defense effected by cultivar-specific resistance – race-non-specific (horizontal resistance) as well as race-specific (vertical resistance).

In agricultural practice SAR can be induced by infecting the cultivar to be protected, employing as the first infecting pathogen a race behaving either avirulently or virulently, but in either case the infection response must produce a sufficiently large necrosis. Alternatively, plants may be sprayed with either culture filtrates of saprophytic gram-positive or gram-negative bacteria (Schönbeck et al. 1980) or, even better, with one of the chemicals identified as signal substances, such as salicylic acid or abscisic acid. Since these substances are easily decomposed biologically and the spectrum of pathogens that can be repelled is very broad, their application to release SAR has some potential for broad spectrum plant protection. Since SAR is less efficient in protecting against one particular pathogen than a mutation to cultivar-specific resistance, further measures may be taken to compensate for this drawback. The intense and ongoing research into SAR, in particular its molecular genetics, will soon show how SAR may be applied successfully, perhaps by combining it with other plant protection measures.

Reviews

Chen Z.X., Malamy J., Henning J., Conrath U., Sanchezcasas P., Silva H., Ricigliano J., Klessig D.F. (1995): Induction, modification, and transduction of the salicylic acid signal in plant defense responses. Proc.Natl.Acad.Sci.USA 92: 4134–4137
Day St. (1993): A shot in the arm of plants. New Scientist 9 January: 36–40
De Wit P.J.G.M. (1985): Induced resistance to fungal and bacterial diseases. In: Fraser R.S.S. (ed.): Mechanisms of Resistance to Plant Diseases. Martinus Nijhoff/Dr.W.Junk, Dordrecht, Boston, Lancaster. 405–424
Dixon R.A., Lamb C.J. (1990): Molecular communication in interactions between plants and microbial pathogens. Annu.Rev.Plant Physiol. 41: 339–367
Dong X. (1998): SA, JA, ethylene, and disease resistance in plants. Curr. Opinion Plant Biol. 1: 316–323
Hammerschmidt R. (1993): The nature and generation of systemic signals induced by pathogens, arthropod herbivores, and wounds. Advances in Plant Pathology 10: 307–337

Hammond-Kosack K.E., Jones J.D.G. (1996): Resistance Gene-Dependent Plant Defense Responses. Plant Cell 8: 1773–1791

Heath M.C. (1995): Thoughts on the role and evolution of induced resistance in natural ecosystems, and its relationship to other types of plant defenses against disease. In: Hammerschmidt R., Kuc J. (eds.): Induced Resistance to Disease in Plants. Kluver Academic Publishers, Dordrecht, Boston, London. 141–151

Hunt M.D., Neuenschwander U.H., Delaney T.P., Weymann K.B., Friedrich L.B., Lawton K.A., Steiner H.-J., Ryals J.A. (1996): Recent advances in systemic resistance research – a review. Gene 179: 89–95

Kuc J. (1983): Induced systemic resistance in plants to diseases caused by fungi and bacteria. In: Bailey J.A., Deverall B.J. (eds.): The Dynamics of Host Resistance. Academic Press Australia, Sidney, New York, London, Paris, San Diego, San Francisco, Sao Paulo, Tokyo, Toronto. 191–221

Kuc J. (1995): Induced systemic resistance – an overview. In: Hammerschmidt R., Kuc J. (eds.): Induced Resistance to Disease in Plants. Kluver Academic Publishers, Dordrecht, Boston, London. 169–175

Lamb C.J. (1994): Plant disease resistance genes in signal perception and transduction. Cell 76: 419–422

Lee H.I., Leon J., Raskin I. (1995): Biosynthesis and metabolism of salicylic acid. Proc.Natl. Acad.Sci.USA 92: 4076–4079

Linthorst H.J.M. (1991): Pathogenesis-related proteins of plants. Crit. Rev. Plant Sci. 10: 123–150

Madamanchi N.R., Kuc J. (1991): Induced systemic resistance in plants. In: Cole G.T., Hoch H.C. (eds.): The Fungal Spore and Disease Initiation in Plants and Animals. Plenum Press, New York, London. 347–362

Malamy J., Klessig D.F. (1992): Salicylic acid and plant disease resistance. Plant J. 2: 643–654

Moffat A.S. (1992): Improving plant disease resistance. Science 257: 482–483

Pena-Cortes H., Fisahn J., Willmitzer L. (1995): Signals involved in wound-induced proteinase inhibitor II gene expression in tomato and potato plants. Proc.Natl.Acad.Sci. USA 92: 4106–4113

Roberts K. (1992): Potential awareness of plants. Nature 360: 14–15

Ryals J., Lawton K.A., Delaney T.P., Friedrich L., Kessmann H., Neuenschwander U., Uknes S., Vernooij B., Weymann K. (1995): Signal transduction in systemic acquired resistance. Proc.Natl.Acad.Sci.USA 92: 4202–4205

Ryals J.A., Neuenschwander U.H., Willits M.G., Molina A., Steiner H.-Y., Hunt M.D. (1996): Systemic Acquired Resistance. Plant Cell 8: 1809–1819

Ryan C.A. (1992): The search for proteinase inhibitor-inducing factor, PIIF. Plant Mol.Biol. 19: 123–133

Simons P. (1992): The secret feelings of plants. New Scientist 17 October: 29–33

Stermer B.A. (1995): Molecular regulation of systemic induced resistance. In: Hammerschmidt R., Kuc J. (eds.): Induced Resistance to Disease in Plants. Kluver Academic Publishers, Dordrecht, Boston, London. 111–140

Vernooij B., Uknes S., Ward E., Ryals J. (1994): Salicylic acid as a signal molecule in plant-pathogen interactions. Curr.Opinion Cell. Biol. 6: 275–279

Relevant papers

Chen Z., Silva H., Klessig D.F. (1993): Active oxygen species in the induction of plant systemic acquired resistance by salicylic acid. Science 262: 1883–1886

Delaney T.P., Uknes S., Vernooij B., Friedrich L., Weymann K., Negrotto D., Gaffney T., Gutrella M., Kessmann H., Ward E., Ryals J. (1994): A central role of salicylic acid in plant disease resistance. Science 266: 1247–1250

Farmer E.E., Ryan C.A. (1990): Interplant communication: Airborne methyl jasmonate induces synthesis of proteinase inhibitors in plant leaves. Proc.Natl.Acad.Sci.USA 87: 7713–7716

Gaffney T., Friedrich L., Vernooij B., Negrotto D., Nye G., Uknes S., Ward E., Kessmann H., Ryals J. (1993): Requirement of Salicylic Acid for the Induction of Systemic Acquired Resistance. Science **261**: 754–756

Hennig J., Malamy J., Grynkiewicz G., Indulski J., Klessig D.F. (1993): Interconversion of the salicylic acid signal and its glucoside in tobacco. Plant J. **4**: 593–600

Metraux J.P., Ahl Goy P., Staub T., Speich J., Steinemann A., Ryals J., Ward E. (1991): Induced systemic resistance in cucumber in response to 2,6-dichloro-isonicotinic acid and pathogenesis. In: Hennecke H., Verma D.P.S. (eds.): Adcances in Molecular Genetics of Plant-Microbe Interactions. Kluwer Academic Publishers, Dordrecht, Netherlands. 432–439

Schönbeck F., Dehne H.-W., Beicht W. (1980): Untersuchungen zur Aktivierung unspezifischer Resistenzmechanismen in Pflanzen. Ztschr. Pflanzenkrankheiten u. Pflanzenschutz **87**: 654–666

Shah J., Klessig D.F. (1996): Identification of a salicylic acid-responsive elementin the promotor of the tobacco pathogenesis-related β-1,3–glucanase gene, *PR-2d*. Plant J. **10**: 1089–1101

Shulaev V., Leon J., Raskin I. (1995): Is salicylic acid a translocated signal of systemic acquired resistance in tobacco? Plant Cell **7**: 1691–1701

Uknes S., Mauch-Mani B., Moyer M., Potter S., Williams S., Dincher S., Chandler D., Slusarenko A., Ward E., Ryals J. (1992): Acquired resistance in *Arabidopsis*. Plant Cell **4**: 645–656

Vernooij B., Friedrich L., Morse A., Reist R., Kolditzjawhar R., Ward E., Uknes S., Kessmann H., Ryals J. (1994): Salicylic acid is not the translocated signal responsible for inducing systemic acquired resistance but is required in signal transduction. Plant Cell **6**: 959–965

Ward E.R., Uknes S.J., Williams S.C., Dincher S.S., Wiederhold D.L., Alexander D.C., Ahl-Goy P., Métraux J.P., Ryals A. (1991): Coordinate gene activity in response to agents that induce systemic acquired resistance. Plant Cell **3**: 1085–1094

Wildon D.C., Thain J.F., Minchin P.E.H., Gubb I.R., Reilly A.J., Skipper Y.D., Doherty H.M., O'Donnell P.J., Bowles D.J. (1992): Electrical signaling and systemic proteinase inhibitor induction in wounded plant. Nature **360**: 62–65

12 Evolution of Plant-Pathogen Interaction – Its Influence on Breeding of Disease Resistant Crops in Agriculture

The ability of plants to defend themselves against parasitism and the potential of pathogens to attack and colonize plants are subject to evolution based on mutation and recombination. The diversity of its many steps and the resulting selections effect a coevolution of plants with their parasites. Only some of the component steps proceede before our eyes in agriculture, such as the coevolution between race-specific resistant cultivars and their pathogens which result from plant breeding (see also Sect. 9.1.5).

In nature, the evolution of parasitism and host plant resistance is also continuing, but its steps are much more difficult to observe, since the events are rarely amplified sufficiently by selection processes to make them recognizable. The genetic alterations they are based on in the pathogen consist of the acquisition, or the loss, of the ability to overcome, bypass, or avoid the plant's defense barriers. In the plant these genetic alterations affect, gain, or lose the capacity to express basic resistance, or cultivar-specific resistance, or the ability to recognize an elicitor or a host-selective toxin. The speed of evolution depends on the frequencies of mutation and recombination, the selective advantage of new genotypes and the number of mutation and/or recombination steps required for a new, fully functional phenotype.

Many mutations may be required for a pathogen to develop new pathogenicity genes for overcoming the basic resistance of a particular plant species or for a host plant to establish a new basic resistance defense mechanism of the species-non-specific resistance type. Such changes are expected to occur only rarely.

On the other side, the loss in the plant of race-non-specific resistance in a short space of time is also an improbable event because of its polygenic determination. Many different genes must become defective before this type of resistance type is lost completely. Accordingly, loss of race-non-specific resistance via natural evolution, i.e., without human interference by the application of particular breeding strategies (see also Sect. 9.2), is not expected to occur.

However, evolution of race-specific resistance of a plant may proceed more rapidly because of its particular genetic determination: Each loss of race-specific resistance may occur either by mutation in the plant gene determining specific recognition of a pathogen race or by mutation to specific virulence in the pathogen's corresponding avirulence gene (see Sect. 9.1 and thereafter). Each mutation in the host and pathogen prevents recognition between the corresponding genes or gene products. The survival of new plant and pathogen genotypes will also depend on their growth rates, generation times, and the number of progeny produced per generation.

As a rule, in natural ecosystems an equilibrium is observed between host plants and their pathogens, i.e., there is some coexistence between both partners safeguarding survival of both. In such natural systems, infection-susceptible plants and plants exhibiting race-specific or race-non-specific resistance exist side by side. In the pathogen population, several different races of each *forma specialis* are also present. But quite contrary to this balanced natural system, plant production in agriculture is running far away from an equilibrium. Cultivation of genetically homogenous, race-specific resistant cultivars in single-crop farming represents an entirely unnatural condition which exerts extreme pressure for selection of specific virulent pathogen races (see also Sect. 9.1.5).

Responsible plant protection measures to fight such unwanted selection pressure must include the breeding and cultivation of pathogen-resistant cultivars refractory as ever possible to loss during long-term use. Several different strategies can be used to pursue this aim: growing (1) race-non-specific resistant cultivars (see Sect. 9.2), (2) infection-tolerant cultivars, but paying attention to the restriction mentioned in Chap. 10, (3) using multiline varieties (see Sect. 9.1.5), (4) using cultivars exhibiting durable resistance which contain genes of different resistance types (see Sect. 9.2), or (5) raising cultivars of high genetic heterogeneity. Furthermore, barriers against pathogen attack or measures against selection of new virulent pathogen races may be improved by construction of new cultivars employing gene technology methods and/or by introducing-crossing from other plant species genetic determinants belonging to basic resistance. Such newly constructed defense mechanisms in plants could be overcome by previously successful pathogens only after the latter have passed through many steps of mutation or recombination effecting development of quite new pathogenicity genes. Finally, the application of induced resistance and SAR in combination with any other of the above mentioned breeding techniques may considerably improve plant protection.

Reviews

Heath M.C. (1987): Evolution of plant resistance and susceptibility to fungal invaders. Can. J. Plant Pathol. **9**: 389–397

Heath M.C. (1991): Evolution of resistance to fungal parasitism in natural ecosystems. New Phytol. **119**: 331–343

Pryor T. (1987): The origin and structure of fungal disease resistance genes in plants. Trends Genet. **3**: 157–161

13 Glossary: Explanation of some Terms Utilized in Biology and Plant Pathology

aggressiveness: Extent of disease production during the interaction of a genetically defined pathogen species or race with particular cultivars. The aggressiveness of a particular pathogen race is determined by the genotypes of the pathogen and the host plant, but may be modified by environmental conditions. Under constant environmental conditions aggressiveness is a reproducible quantitative measure. (See in contrast → parasitic fitness.)

allele: A particular form of a gene. Different alleles of one gene belong to the same locus on a chromosome. An allele may be changed into another by mutation. All the alleles of a single gene belong to the same biochemical or ontogenetic process. – If two or more alleles with different phenotypes map at the same locus they are called → multiple alleles.

avirulence: The failure of an infecting homologous pathogen carrying an avirulence gene to parasitize a race-specific resistant host plant carrying a "corresponding" race-specific resistance gene.

axenic cultivation: The cultivation of an organism in the absence of any other organisms.

basic incompatibility: See basic resistance.

basic compatibility: Compatibility between a pathogen and its host plant that results in parasitism of the latter (see also compatible interaction).

basic resistance (non-host resistance, basic incompatibility, multigene resistance): The resistance of a non-host plant that is directed against many different pathogens. Basic resistance implies the absence of → basic compatibility. Basic resistance is observed if the pathogen lacks pathogenicity genes able to overcome the plant's defense mechanisms.

biotrophy: A mode of nutrition requiring living and metabolizing plant cells as a source of nutrients. Some biotrophic pathogens may be grown axenically in culture media supplemented with the chemicals otherwise obtained from living plant cells. Obligate biotrophs, can be grown only in living and metabolizing plants. (See also facultative biotrophy.)

codominance: Independent expression of alleles.

compatible interaction: An interaction between a susceptible host plant and a pathogen that results in unrestricted parasitism of the plant. (See also homologous interaction).

conditional gene expression: Expression of genes only under particular conditions, such as the absence of a suppressor gene or within a defined temperature range.

corresponding resistance and avirulence genes: This term designates genes for race-specific resistance in host plants that only interact with particular avirulence genes in homologous pathogens. Each race-specific resistance gene forms with the avirulence gene that interacts with it a so called corresponding gene pair. The products of both genes interact or "recognize" each other. The recognition reaction results in the expression of defense reactions against the attacking pathogen.

cultivar: Variety of a cultivated plant originating from breeding activities.

cultivar-specific resistance: Resistance of a particular variety of a cultivated plant directed against a particular race or a *forma specialis* of a pathogen. However, not all pathogens have *formae speciales*.

differential resistance: → race-specific resistance

dikaryon: A fungal cell or mycelium made up of binucleate cells. The dikaryotic state may be found among ascomycetes for short periods of their life cycles and much more commonly in basidiomycetes.- The phenotype of a dikaryon is similar to that of a → diploid.

diploid: An organism with two homologous sets of chromosomes. As a rule this results from karyogamy, i.e. sexual fusion of two → haploid nuclei each containing a single (1n) chromosomal set, one of paternal and one of maternal origin.

DNA-restriction fragment: DNA fragment generated by cuts in a larger DNA molecule made with a DNA restriction endonuclease.

DNA restriction pattern: The pattern or distribution of DNA restriction fragments seen after gel electrophoresis.

dominant: Mode of phenotypic expression of genes in → heterozygous state. i.e. in a diploid organism. The expression is dominant, if the gene in heterozygous state exhibits the same phenotype as in the → homozygous one.

epistatic: Expression of a gene effecting either partial or complete suppression of expression of another, non-allelic gene.

facultative biotrophy: Ability of a pathogen to derive nutrition either by → biotrophy or → necrotrophy.

fitness: Ability of a particular genotype to compete successfully within a population with other individuals.

***forma specialis* (f.sp.):** A subspecies of a pathogen that is specialized on, or confined to, one particular host plant species. Parasitism of the host plant is dependent on the expression of → pathogenicity genes that are specifically directed to the colonization of that particular host.

generalized resistance: → race-non-specific resistance.

genetic polymorphism: The presence in a population of two or more alleles at the same locus

genotype: Hereditary constitution of an individual organism or of nuclei within its cells.

haploid: An organism or nucleus with a single set of chromosomes (1n).

haustorium: Part of a fungal hypha that penetrates the host cell wall and takes up nutrients from the living plant cell. Haustoria are completely surrounded by the plant's plasma membrane, and do not penetrate it. They are located in the periplasmic space, not inside the protoplast.

hemibiotrophy: Mode of nutrition of phytopathogenic fungi employing first → biotrophy, and subsequently → necrotrophy. Also called → facultative bio-trophy.

heterogenote: Designation for a → haploid cell with a duplicated chromosome segment carrying one or several genes with different → alleles.

heterokaryon: A multinucleate cell with nuclei of more than one genotype.

heterologous hypersensitive reaction: A hypersensitive reaction induced by a pathogen attacking a non-host plant. May also be called → heterologous in-compatibility between plant and pathogen.

heterologous interaction: Interaction between a non-host plant and a pathogen. This may result in a → hypersensitive reaction, in this case a → heterologous hypersensitive reaction.

heterothallism: A genetic mechanism restricting sexual fusion in fungi to mycelia of different mating types, often called "+" and "−".

heterozygote: a → diploid cell carrying different → alleles for one or more genes in homologous chromosome segments.

homologous incompatibility: Incompatibility between a pathogen and its host plant. This implies that the host resists the → homologous pathogen in spite of its → basic compatibility because the host plant carries a gene(s) for → race-specific resistance which establishes a secondary defense mechanism active only against particular pathogen races. Race-specific resistance depends on a gene-for-gene interaction between the gene products of host and pathogen.

homologous interaction: Interaction between a host plant and a pathogen that has evolved to parasitise that host. A homologous interaction may result in → compatible interaction, i.e. in parasitism of the host plant, or, if the plant carries a gene for → race-specific resistance, an → incompatible interaction, i.e. resistance.

homologous pathogen: Pathogen that carries → pathogenicity genes that "fit" the host plant.

homologous recombination: Genetic exchange by crossing over between two DNA molecules of identical or nearly identical base sequences.

homozygote: A → diploid cell carrying identical → alleles for one or more genes in homologous chromosome segments.

horizontal resistance: → race-non-specific resistance.

host plant: Plant species that can be parasitized by a particular pathogen.

host range: The group of host plant species that can be parasitized by a particular pathogen.

host recognition: The event that occurs early in the contact between pathogen and plant that determines whether the plant will promote or prevent growth of the attacking pathogen. In molecular terms recognition is a specific interaction between a molecule, or molecular complex, of the pathogen with another of the plant that releases or triggers further reactions.

host resistance: Defense response of a host plant against a homologous pathogen. Also called → cultivar-specific resistance, of which two types are known: → race-specific or → vertical resistance, and → race-non-specific or → horizontal resistance.

HR: → hypersensitive reaction

hypersensitive reaction: An immediate defense reaction of a race-specific resistant plant triggered by an attacking pathogen. Can also be induced by abiotic wounding. The reaction released in the plant may be recognized within a few minutes up to several hours and results in rapid killing of the afflicted plant cell(s). The hypersensitive killing is mostly accompanied by necrotization of immediately adjacent tissue forming a macroscopic local necrosis. The plant "sacrifices", so to speak, its own impaired tissue to contain the pathogen. With race-specific resistant host plants HR is the most common defense reaction against → biotrophic and → hemibiotrophic pathogens. However, HR has also been observed in some non-host plants interacting with pathogens. In this case one speaks of an → heterologous HR.

illegitimate recombination: Genetic exchange and recombination, respectively, between two non-homologous DNA molecules with different base sequences. Illegitimate recombination causes deletions, duplications or other rearrangements in DNA molecules or chromosomes.

incompatible interaction: Interaction of a → race-specific resistant host plant with a homologous pathogen resulting in a defense reaction against the pathogen. Also designated as → incompatibility between both partners (see also homologous interaction).

karyogamy: Fusion of nuclei of two gametes of opposite sex.

monogenic: Determined by a single gene.

multigenic: Determined by several genes.

multigene family: Set of genes developed during evolution by duplication of a single gene. All members of the gene family exhibit extensive homology in their nucleotide sequences and may encode functionally identical, related, or even dissimilar proteins. Single members of a gene family may be expressed during development at different times or in response to different external stimuli.

multigene resistance: Generally used as synonym for → basic resistance. However, non-race specific resistance is also frequently multigenic.

multiple allelism: The occurrence of more than two alleles at the same chromosomal locus, where each allele has a distinct phenotype.

necrotrophy: A mode of nutrition utilizing materials derived from dead plant material.

non-host resistance: → basic resistance

obligate biotrophy: Mode of nutrition employed by fungal parasites that live exclusively on living and fully metabolizing plant cells. So far, most obligate biotrophs can not (yet) be cultivated in vitro, i.e. outside living cells in axenic culture.

oligogenic: A phenotype determined by only a few genes. Oligogenic and monogenic phenotypes are also called qualitative genetic traits that depend on the expression of "major genes".

parasite: Organism living on, colonizing, and reproducing in a host organism. A parasite takes metabolites from its host organism for its own nutrition and reproduction giving nothing in return to the host.

parasite-specific resistance: → cultivar-specific resistance.

parasitic fitness: The ability of a particular pathogen genotype to survive and propagate itself in the pool of pathogens that already exist in a population of host plants.

pathogen: Parasite causing disease symptoms in its host plant.

pathogenicity: Ability of a parasite to cause disease symptoms. Pathogenicity is a → qualitative character, i.e. the parasite can be either pathogenic or nonpathogenic.

pathogenicity factor: Product of a pathogenicity gene of a parasite that contributes to parasitism and hence the expression of a plant disease.

pathogenicity gene: Genes that determine the activities of a parasite when attacking a particular plant that enable the parasite to overcome the plant's → basic resistance.

perthotrophy: A mode of nutrition in which the pathogen first kills the host tissue by the action of enzyme(s) or toxin(s) and then exploits the dead plant material. The majority of phytopathogenic fungi and bacteria adopt this life style.

phenotype: The appearance or properties of an organism determined by the interaction of its genotype with the environment.

physiological race: Pathogen races that are distinguished by their reactions on a set of differential hosts carrying different race specific genes for resistance.

phytoalexin: A water soluble and antibiotic active substance of low molecular weight synthesized by race-specific resistant host plants after pathogen attack or after wounding or other stress.

phytotoxin: Toxin synthesized by a phytopathogen that affects plants. Two types can be discriminated; host-selective phytotoxins act only on host plants which recognize the toxin by a specific receptor protein; host-non-selective toxins are general nonspecific cell poisons.

plasmid: An extrachromosal DNA element capable of independent replication. Commonly found in prokaryotes, but seldom in eukaryotes. Plasmids are much smaller than chromosomes existing in one or several copies per cell of single- or double stranded, circular or linear DNA molecules. Generally, plasmids do not carry genes essential for cell survival so their loss is usually not lethal.

plasmogamy: Fusion of protoplasts of two cells of opposite sex without accompanying fusion of their nuclei.

polygenic: Genetic determination of an organism's character by the additive action of many "minor genes". Each one of the minor genes delivers a small contribution to the observed character. Polygenic traits are called → quantitative characters.

polymorphism: → genetic polymorphism.

polyploid: Designation for a cell with three or more homologous chromosome sets.

qualitative character: A trait usually determined by one or a few genes.

quantitative character: A trait usually determined by many genes with individually small effects.

race-non-specific resistance: A genetically stable, → polygenically determined resistance directed against all → physiological races of a homologous pathogen. Also called → horizontal resistance, → uniform resistance, → generalized resistance.

race-specific resistance: A rather unstable resistance directed only against single races of a pathogen. It depends on expression of different defense mechanisms, and is mostly directed against biotrophic pathogens. Depends on a gene-for-gene interaction between the products of a resistance gene in the plant and a "corresponding" avirulence gene in the pathogen. The genetic instability of race-specific resistance results from the high mutation rate among the corresponding pathogen avirulence genes to → specific virulence.

recessive: Not expressed in the → heterozygous state.

saprophyte: An organism that feeds on dead material.

specific resistance: → race-specific resistance.

specific virulence: Virulent behavior of a pathogen which lacks an avirulence allele which would otherwise lead to its recognition by a host plant harboring a → race-specific resistant gene. As a consequence no defense reactions are released in the host plant despite the presence of the race-specific resistance gene.

symbiosis: Living together of two different organisms which benefit equally from their mutual metabolic dependence.

systemic spreading: Spreading of a substance or pathogen throughout the plant.

uniform resistance: → race-non-specific resistance.

vertical resistance: → race-specific resistance.

virulence: The ability of a pathogen to produce disease on a race specific resistant host. (See also → avirulence and → specific virulence). The term virulence is also used for toxin forming pathogens to describe the intensity of toxin action.

zygote: Diploid cell arising from the fusion of two haploid gametes.

Subject Index

Keywords with page numbers in **bold** are explained in the Glossary (Chap. 13).

Printing and Binding: Stürtz AG, Würzburg